Advances in Intelligent Systems and Computing

Volume 293

Series editor

Janusz Kacprzyk, Polish Academy of Sciences, Warsaw, Poland
e-mail: kacprzyk@ibspan.waw.pl

For further volumes:
http://www.springer.com/series/11156

About this Series

The series "Advances in Intelligent Systems and Computing" contains publications on theory, applications, and design methods of Intelligent Systems and Intelligent Computing. Virtually all disciplines such as engineering, natural sciences, computer and information science, ICT, economics, business, e-commerce, environment, healthcare, life science are covered. The list of topics spans all the areas of modern intelligent systems and computing.

The publications within "Advances in Intelligent Systems and Computing" are primarily textbooks and proceedings of important conferences, symposia and congresses. They cover significant recent developments in the field, both of a foundational and applicable character. An important characteristic feature of the series is the short publication time and world-wide distribution. This permits a rapid and broad dissemination of research results.

Javier Bajo Perez · Juan M. Corchado Rodríguez
Philippe Mathieu · Andrew Campbell
Alfonso Ortega · Emmanuel Adam
Elena M. Navarro · Sebastian Ahrndt
María N. Moreno · Vicente Julián
Editors

Trends in Practical Applications of Heterogeneous Multi-agent Systems. The PAAMS Collection

Springer

Editors
Javier Bajo Perez
Department of Artificial Intelligence
Technical University of Madrid
Salamanca, Spain

Juan M. Corchado Rodríguez
Faculty of Science
Dept. of Computing Science and Control
University of Salamanca
Salamanca, Spain

Philippe Mathieu
Lille University of Science and Technology
Villeneuve dŠAscq, France

Andrew Campbell
Dartmouth College
Hanover, New Hampshire, USA

Alfonso Ortega
Autonomous University of Madrid
Madrid, Spain

Emmanuel Adam
University of Valenciennes
Valenciennes, France

Elena M. Navarro
University of Castilla-La Mancha
Albacete, Spain

Sebastian Ahrndt
Distributed Artificial Intelligence Lab
Fakultät IV Electrical Engineering and Computer Science
Technical University of Berlin
Berlin, Germany

María N. Moreno
Universidad de Salamanca
Salamanca, Spain

Vicente Julián
Polytechnic University of Valencia
Valencia, Spain

ISSN 2194-5357 ISSN 2194-5365 (electronic)
ISBN 978-3-319-07475-7 ISBN 978-3-319-07476-4 (eBook)
DOI 10.1007/978-3-319-07476-4
Springer Cham Heidelberg New York Dordrecht London

Library of Congress Control Number: 2014939944

Printed on acid-free paper

Springer is part of Springer Science+Business Media (www.springer.com)

Preface

PAAMS'14 Special Sessions are a very useful tool in order to complement the regular program with new or emerging topics of particular interest to the participating community. Special Sessions that emphasized on multi-disciplinary and transversal aspects, as well as cutting-edge topics were especially encouraged and welcome.

Research on Agents and Multi-Agent Systems has matured during the last decade and many effective applications of this technology are now deployed. An international forum to present and discuss the latest scientific developments and their effective applications, to assess the impact of the approach, and to facilitate technology transfer, has become a necessity.

PAAMS, the International Conference on Practical Applications of Agents and Multi-Agent Systems is an evolution of the International Workshop on Practical Applications of Agents and Multi-Agent Systems. PAAMS is an international yearly tribune to present, to discuss, and to disseminate the latest developments and the most important outcomes related to real-world applications. It provides a unique opportunity to bring multi-disciplinary experts, academics and practitioners together to exchange their experience in the development of Agents and Multi-Agent Systems.

This volume presents the papers that have been accepted for the 2014 edition in the special sessions: Agents Behaviours and Artificial Markets (ABAM), Agents and Mobile Devices (AM), Bio-Inspired and Multi-Agents Systems: Applications to Languages (BioMAS), Multi-Agent Systems and Ambient Intelligence (MASMAI), Self-Explaining Agents (SEA), Web Mining and Recommender systems (WebMiRes) and Intelligent Educational Systems (SSIES).

We would like to thank all the contributing authors, as well as the members of the Program Committees of the Special Sessions and the Organizing Committee for their hard and highly valuable work. Their work has helped to contribute to the success of the PAAMS'14 event. Thanks for your help, PAAMS'14 wouldn't exist without your contribution.

We thank the sponsors (IBM, JCyL, IEEE Systems Man and Cybernetics Society Spain, AEPIA Asociación Española para la Inteligencia Artificial, AFIA French Association for Artificial Intelligence, CNRS Centre national de la recherche scientifique, AIXIA Associazione Italiana per l'Intelligenza Artificiale), the Local Organization members and the Program Committee members for their hard work, which was essential for the success of PAAMS'14.

Juan Manuel Corchado
Javier Bajo Perez
PAAMS'14 Organizing Co-chairs

Organization

Special Sessions

Agents Behaviours and Artificial Markets
Agents and Mobile Devices
Bio-Inspired and Multi-Agents Systems: Applications to Languages
Multi-Agent Systems and Ambient Intelligence
Self-Explaining Agents
Web Mining and Recommender systems
Intelligent Educational Systems

Special Session on Agents Behaviours and Artificial Markets

Philippe Mathieu (Chair)	University of Lille, France

Scientific Committee

Bruno Beaufils	University of Lille1, France
Olivier Brandouy	University of Paris 1, France
Florian Hauser	Austria, Austria
Philippe Mathieu	University of Lille1, France
Adolfo López Paredes	University of Valladolid, Spain
Javier Arroyo	University Complutense Madrid, Spain
Marco Raberto	University of Genoa, Italy
Roger Waldeck	Telecom Bretagne, France
Murat Yildizoglu	University of Bordeaux IV, France

Special Session on Agents and Mobile Devices

Andrew Campbell	Darthmouth College, USA
Javier Bajo	Polytechnic University of Madrid, Spain

Scientific Committee

Antonio Juan Sánchez	University of Salamanca, Spain
Juan Francisco De Paz	University of Salamanca, Spain
Gabriel Villarrubia	University of Salamanca, Spain
Cristian Pinzón	Technical University of Panama, Panama
Montserrat Mateos	Pontifical University of Salamanca, Spain
Luis Fernando Castillo	University of Caldas, Colombia
Miguel Ángel Sánchez	Indra, Spain
Roberto Berjón	Pontifical University of Salamanca, Spain
Encarnación Beato	Pontifical University of Salamanca, Spain
Fernando De la Prieta	University of Salamanca, Spain

Special Session on Bio-Inspired and Multi-Agents Systems:

Applications to Languages

María Dolores Jiménez López	Universitat Rovira i Virgili, Spain
José Luis Montaña Arnaiz	Universidad de Cantabria, Spain
Cesar Luis Alonso	Universidad de Oviedo, Spain
Alfonso Ortega de la Puente	Universidad Autónoma de Madrid, Spain
Marina De la Cruz Echeandía	Universidad Autónoma de Madrid, Spain

Special Session on Multi-Agent Systems and Ambient Intelligence

Antonio Fernández-Caballero (Co-Chair)	University of Castilla-La Mancha, Spain
Elena María Navarro (Co-Chair)	University of Castilla-La Mancha, Spain

Scientific Comittee

Javier Jaén-Martínez	Universidad Politécnica de Valencia, Spain
Mª Teresa López-Bona	University of Castilla-La Mancha, Spain
Diego López-de-Ipiña	University of Deusto, Spain
Rafael Martínez-Tomás	National Distance Education University, Spain
Paulo Novais	Universidade do Minho, Spain
Juan Pavón	University Complutense Madrid, Spain

Special Session on Self-Explaining Agents

Sebastian Ahrndt (Co-Chair)	Technische Universität Berlin, Germany
Johannes Fähndrich (Co-Chair)	Technische Universität Berlin, Germany
Benjamin Hirsch (Co-Chair)	Khalifa University, United Arab Emirates

Program Committee

Sebastian Ahrndt	DAI-Labor, Technische Universität Berlin, Germany
Johannes Fähndrich	DAI-Labor, Technische Universität Berlin, Germany
Benjamin Hirsch	EBTIC, Khalifa University, United Arab Emirates
Marco Lützenberger	DAI-Labor, Technische Universität Berlin, Germany
Arkady Zaslavsky	CSIRO, Australian National University, Australia
Michael Kaisers	Centrum Wiskunde & Informatica, Netherlands

Special Session on Web Mining and Recommender systems

María Moreno (Chair)	University of Salamanca, Spain

Scientific Committee

Ana María Almeida	Institute of Engineering of Porto, Portugal
Yolanda Blanco Fernández	University of Vigo, Spain
Rafael Corchuelo	University of Sevilla, Spain
Chris Cornelis	Ghent University, Belgium
María José del Jesús Díaz	University of Jaén, Spain
Anne Laurent	University of Montpellier 2, France
Vivian López Batista	University of Salamanca, Spain
Joel Pinho Lucas	Mobjoy, Brazil
Constantino Martins	Institute of Engineering of Porto, Portugal
Yong Zheng	DePaul University, USA

Special Session on Intelligent Educational Systems

Juan Miguel Alberola (Co-Chair)	Universitat Politecnica de Valencia, Spain
Elena Del Val (Co-Chair)	Universitat Politecnica de Valencia, Spain
Vicente Julian (Co-Chair)	Universitat Politecnica de Valencia, Spain
Vicente Botti (Co-Chair)	Universitat Politecnica de Valencia, Spain

Scientific Committee

Juan Miguel Alberola	Universitat Politecnica de Valencia, Spain
Elena Del Val	Universitat Politecnica de Valencia, Spain
Vicente Julian	Universitat Politecnica de Valencia, Spain
Vicente Botti	Universitat Politecnica de Valencia, Spain
Ricardo Azambuja Silveira	Universidade Federal de Santa Catarina, Brazil
José Cascalho	Universidade dos Açores, Portugal
César Alberto Collazos Ordóñez	Universidad del Cauca, Colombia
Helder Manuel Ferreira Coelho	Universidade de Lisboa, Portugal
Francisco José García Peñalvo	Universidad de Salamanca, Spain
Carina Soledad González	Universidad de la Laguna, Spain
Blair Lehman	University of Memphis, USA
Tingting Li	University of Bath, United Kingdom
Antonija Mitrovic	Comp. Science and Software Engineering, New Zealand
Paulo Novais	Universidad do Minho, Portugal
Andrea Omicini	Alma Mater Bologna, Italy
Carlos Vaz de Carvalho	Instituto politécnico do Porto, Portugal
Laurent Vercouter	INSA de Rouen, France

Contents

Normal Distributions and Multi-issue Negotiation for Service Composition*

Silvia Rossi[1], Dario Di Nocera[2,**], and Claudia Di Napoli[3]

[1] Dipartimento di Ingegneria Elettrica e Tecnologie dell'Informazione
University of Naples "Federico II", Napoli, Italy
silvia.rossi@unina.it
[2] Dipartimento di Matematica
University of Naples "Federico II", Napoli, Italy
dario.dinocera@unina.it
[3] Istituto di Calcolo e Reti ad Alte Prestazioni
C.N.R., Napoli, Italy
claudia.dinapoli@cnr.it

Abstract. Software negotiation is gaining an increased popularity as a viable approach to establish agreements between service providers and consumers of QoS-aware Service-Based Applications (SBA) composed of services provided by different agents. In most cases, QoS preferences are expressed as end-to-end quality requirements on the whole application, and different service agents have to provide services with QoS values that, once aggregated, have to meet them. In the present work we analyze the properties of a hybrid iterative negotiation mechanism occurring among a composer agent and service provider agents on the QoS attributes of the required SBA. The proposed negotiation relies on normal probability distributions to model service provider agents, and it allows to model single-issue and multi-issue negotiation within the same negotiation framework in terms of adopted concession strategy, utility and protocol.

1 Introduction

The increased popularity of Service Oriented Computing [1] is enhancing the development of Service Based Applications (SBAs), i.e. distributed applications composed of services provided by independent and autonomous providers, in a loosely coupled manner that collectively fulfill a requested task. Usually, users requesting an SBA specify also non-functional requirements, referring to Quality

* The research leading to these results has received funding from the EU FP7-ICT-2012-8 under the MIDAS Project (Model and Inference Driven - Automated testing of Services architectures), Grant Agreement no. 318786, and the Italian Ministry of University and Research and EU under the PON OR.C.HE.S.T.R.A. project (ORganization of Cultural HEritage for Smart Tourism and Real-time Accessibility).
** Ph.D. scholarship funded by Media Motive S.r.l., Napoli - Italy, POR Campania FSE 2007-2013.

J.B. Perez et al. (eds.), *Trends in Practical Applications of Heterogeneous Multi-agent Systems. The PAAMS Collection,* Advances in Intelligent Systems and Computing 293,
DOI: 10.1007/978-3-319-07476-4_1, © Springer International Publishing Switzerland 2014

of Service (QoS) attributes of the application, that need to be fulfilled by the providers of the component services. Typical examples of QoS attributes are price, response time, reliability, reputation, and so on. Users may not need to be part of the composition process as long as the functional and non-functional requirements they specify are satisfied [2]. Negotiation mechanisms were shown to be a suitable approach to deal with QoS-aware SBAs [2,3,4], allowing to create composition of services that meet the users QoS requirements. Negotiation occurs on the QoS attributes of the services composing the application that represent the issues of negotiation.

However, practical negotiation mechanisms for SBA applications must be computationally efficient [4], and negotiation strategies should be developed based on the assumption of bounded rather than perfect rationality [5]. Moreover, negotiation protocols for these applications have to be more complex than traditional bilateral negotiation. In most cases, when dealing with QoS-aware SBAs, there are multiple QoS attributes representing non-functional characteristics of the SBA component services, so negotiation has to be modeled as a multi-issue one. While one-issue negotiation is widely studied in literature, negotiation on multiple-issue is less mature [6]. Moreover, the tractability requirement is fundamental in order to apply negotiation mechanisms when services are made available on the market to end users with QoS values depending on market trends. In such settings, service applications providers and consumers have to engage in interactions easy to model and that quickly converge to an agreement.

In this paper we show that negotiation in SBAs is inherently multi-issue even in the case of a single issue negotiation. In fact, also in the case of one QoS attribute for each service composing the application, the QoS value of the complete application, that have to meet the end-to-end constraint required by the user, is obtained by composing the single QoS values provided by the different component services. In particular, we show that when dealing with composition of services, the same utility functions and strategies used to model a negotiation on r issues, representing the QoS attributes of a single service, can be used for the case of one-issue negotiation for r services composing the application. This means that the complexity of the negotiation for SBAs depends on the number of issues and on the number of services composing the SBA in an uniform way. Finally, we show that non linear utility functions, as well as concession strategies, can be modeled through the use of normal distributions with an uniform strategy approach. Normal distributions convolution properties when scaling up in dimensions, allow to deal with computational tractability requirements necessary in real market of services. These properties allow to use the same negotiation mechanism in terms of protocols, strategies and utilities for both single issue and multi-issue negotiation, when dealing with composition of services. Furthermore, the use of normal distributions allows to simulate the stochastic behaviour of service providers with zero-intelligence that can be used to approximate the trends of a volatile and open market of services [7]. Stochastic behavior can be often observed in practical multi-agent negotiation applications [8].

2 One-to-Many-to-Many Asymmetric Negotiation

In many real market situations, provider agents may adopt negotiation strategies to formulate offers, while composer may only have range of acceptance (in some case flexible) on the complete set of offers for the SBA QoS attributes they require. They are not able to provide single counteroffers for each service, but they can only evaluate acceptable and unacceptable offers for the complete package. So, symmetric protocols requiring a strong symmetry between composers and providers [9] are not appropriate.

In [3], we proposed a *one-to-many-to-many* negotiation mechanism allowing only the provider agents to formulate new offers for the issue to be negotiated upon, and only the composer agent (acting on behalf of the user) to evaluate them both individually and globally. The rationale of this choice is that offers for a single functionality cannot be evaluated independently from the ones received for the other functionalities when the value of the issue to be negotiated upon results from the composition of the values provided by the component services. In fact, in such a case, when the value of the issue provided by one services changes, also the values of the same issue provided by the other component services have to change accordingly in order to meet the user's constraint. The protocol of the negotiation is based on an Iterative Contract Net Protocol. It allows a composer agent to negotiate separately with all the agents available for each service, and it may be iterated for a variable number of times (*rounds*) until a *deadline* is reached or the negotiation is successful. A successful negotiation occurs if a complete set of offers, one for each service composing the requested SBA, is accepted.

3 Single and Multi-issue Strategy for the Provider

In order to prepare an offer, an agent uses a set of tactics to generate new values for each negotiated issue [10]. In fact, agents must be provided with strategies to formulate offers and they must be equipped with algorithms to evaluate their utilities for the offers. This is done by evaluating an offer in terms of agent utility with respect to the offer. For each provider, a negotiation strategy on a single parameter (q_1) can be modeled by a Gaussian function [3], as shown in Figure 1a. Such distribution is used both to map values of the single QoS into an utility value for the provider, but also as a strategy to select concession values of the utility. In particular, the Gaussian function represents the probability distribution of the offers in terms of the provider's utility as follows: $U_x(q_1) = \exp\left(-\frac{1}{2}\left(\frac{q_1-\mu_1}{\sigma_1}\right)^2\right)$.

As shown in Figure 1a, the mean value of the Gaussian μ represents the best offer the provider agent may propose in terms of its own utility ($U(\mu) = 1$), but, at the same time, the QoS value with the highest probability to be selected. The standard deviation σ represents the attitude of the provider to concede during negotiation (i.e. greater deviation corresponds to higher concession rate), and the

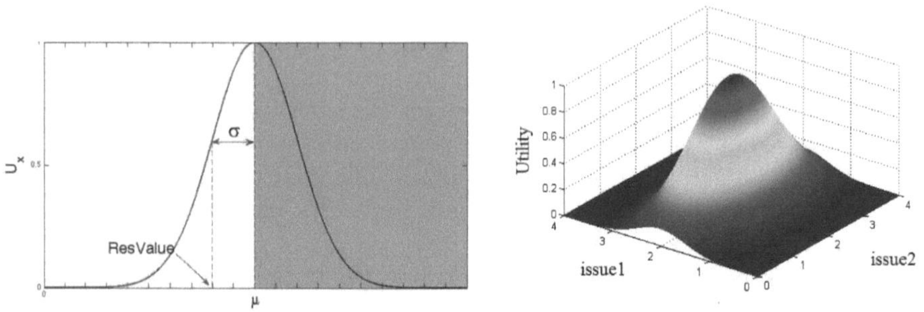

(a) Probability distributions for one-issue. (b) Probability distribution for multi-issues.

Fig. 1. Utilities functions and Gaussian distributions

reservation value (*ResValue*) is set equal to $\mu - \sigma$. In this way the reservation value for a provider is related to its concession strategy, and it is also a value with a known probability to be selected according to its probability distribution. The negotiation set (i.e. the negotiation space for the provider) to be considered is only $[\mu - \sigma; \mu]$ (or $[\mu; \mu + \sigma]$), so only the white (or grey) section of Figure 1a is considered. Values of the utility function are in the domain [0,1], i.e. it is normalized so the values of the QoS attributes are evaluated according to the same scale in order to avoid inaccurate evaluation due to different measurement metrics used for different QoS attributes.

At each negotiation round, a provider generates, following its probability distribution, a new utility value corresponding to a new offer. In order to follow a monotonic concession protocols, if the utility value of this new offer is lower than the one offered in the previous round and within the negotiation set, then the provider proposes the new value. If this value is greater than the one offered in the previous round, or it is outside the negotiation set, the provider proposes the same value offered in the previous round. This strategy allows to simulate different and plausible behaviours of providers that prefers not having a constant loss in utility, even though by increasing the number of negotiation rounds the probability for the provider to move towards its reservation value increases.

The chosen function was shown to be both time and resource dependent. In fact, it takes into account both the *computational load* of an agent, driving its attitude to concede, and the *computational cost* of the provided service corresponding to its best utility value. The computational load of the provider accounts for its workload in terms of the amount of resources it has to provide the service implementations it committed to deliver; while the time dependency is intrinsic in the use of a probabilistic function.

In this work, we show that the negotiation strategy for one-issue negotiation with multiple providers of different services, can be easily extended to multi-issues cases, and that such strategy has relevant properties in the negotiation

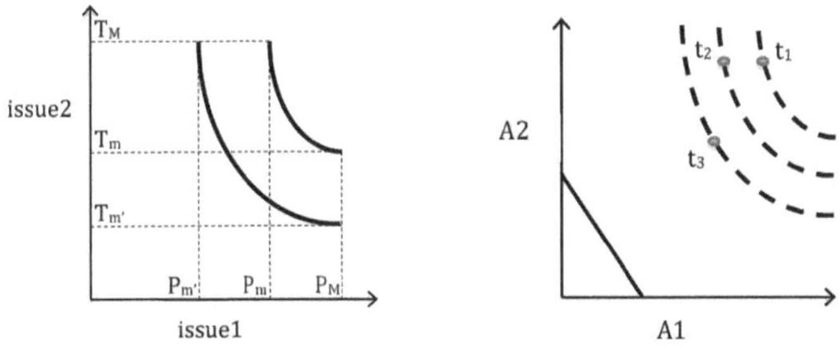

(a) Two-issues indifference curves for different utility values.

(b) One-issue negotiation spaces for a two service composition.

Fig. 2. Negotiation spaces

process. In the multi-issue case, instead of the mono-dimensional Gaussian function used for one-issue negotiation, a multi-dimensional one is adopted. It models both provider's utility, and its attitude to concede by providing offers varying on the Gaussian function. The provider's utility is modeled as follows:

$$U_x(q_1, \ldots, q_r) = \prod_{i=1}^{r} \left[\exp\left(-\frac{1}{2} \left(\frac{q_i - \mu_i}{\sigma_i} \right)^2 \right) \right]$$

where, for each issue q_i, σ_i models the concession attitude when all the other $r-1$ issues are kept fixed, and μ_is are the values for the issues corresponding to the best utility ($U(\mu_1, \ldots, \mu_r) = 1$) for the providers. Values of the utility function are still in the domain [0,1], that is one dimensional (see Figure 1b).

This general representation allows on one hand to model an utility function with non linear dependencies among different issues, but at the same time allows to model a "probabilistic" concession strategy that takes into account different concessions attitudes (σ_i) on different issues. Starting from this multidimensional Gaussian function, an utility level corresponds to an indifference curve, that includes a combination of values, one for each issue, having the same utility value for the provider. In Figure 2a we show different negotiation spaces (section of ellipses) generated for different values of U_x. Such spaces correspond to a negotiation domain that is rational and strictly convex (properties widely applied in economics) [6]. Differently from the one-issue case, here the agent can do tradeoffs between values with the same utility. Tradeoffs in a continuous space may become intractable. Efficient heuristics to find pareto or quasi-pareto optimal solutions exist, but such approaches rely on the availability of counteroffers from the other agents [6].

Moreover, the agent can concede in utility selecting a new negotiation space. When conceding in utility the agent fixes $r-1$ issues and makes a concession on a single issue. Convexity of the utility function ensures that the agent preference on each issue is monotone when fixing the others. So, if the value increases (or decreases), the utility always decreases (or increases). In the multi-issue case, as in the one-issue case, the provider agent generates a new value of utility corresponding to a new offer following its normal distribution. For example, in Figure 2a the rightmost curve represents points for a starting constant value of utility: $U_x(p,t) = const$, with $p \in [P_m, P_M]$ and $t \in [T_m, T_M]$. Fixing $t = T_M$ (for the $issue_2$ in Figure 2a) the agent selects a new value for p (for the $issue_1$ in Figure 2a) from a single issue normal distribution $p = P_{m'}$. To this new set corresponds a new marginal utility value $(U_x(P_{m'}, T_M))$, where $U_x(P_{m'}, T_M) < U_x(P_m, T_M)$.

4 Single Issue Negotiation for Service Composition

In this section we show that the utility functions and strategies used to model a negotiation on r issues for each service of an SBA, are the same as for the case of a single issue negotiation for r services when the value of the single issue of the required SBA is given by the aggregation of the r component values provided by each service of the r services in the SBA. The aggregation function depends on the considered issue. When the issue is additive, as considered in this paper, and each component value is modeled as a normal distribution, such aggregation function is a convolution of their probability distributions.

As an example, here we consider the case of an SBA composed of 2 services $S1$ and $S2$, where the issue under negotiation is the price given by the sum of p_1 and p_2, that are respectively the QoS price values for $S1$ and $S2$, and the end-to-end QoS user's requirement is the global price ($globalPrice$) for the complete application ($p_1 + p_2$). Since the values of the two variables p_1 and p_2 vary according to normal distributions $f_{S1}(p_1)$, $f_{S2}(p_2)$, then the distribution of the variable $z(p_1, p_2) = p_1 + p_2$ is still a normal distribution obtained as the convolution of $f_{S1}(p_1)$ and $f_{S2}(p_2)$:

$$(f_{S1} * f_{S2})(z) = \int f_{S2}(z - p_1) f_{S1}(p_1) dp_1 = \exp\left[-\frac{(z-(\mu_{S1}+\mu_{S2}))^2}{2(\sigma_{S1}^2+\sigma_{S2}^2)}\right]$$

Hence, this convolution can be used to evaluate the distance of the end-to-end QoS requirements from the aggregated QoS values received at a given round as in the case of a multi-issue negotiation with one single service.

In Figure 2b, the corresponding negotiation space is depicted.The dotted curves represent the projection, in a bi-dimensional space, of the section of the Gaussian resulting from the convolution of the two Gaussian functions $f_{S1}(p_1)$ and $f_{S2}(p_2)$ obtained by intersecting the Gaussian with a plane representing a constant utility; the points on the dotted curves represent the aggregated price obtained by a combination of offers received at round t; the line represents the end-to-end QoS constraint, i.e. $p_1 + p_2 = globalPrice$.

The compositor agent adopts the same strategy for both one-issue and multi-issue negotiation evaluating an Euclidean distance between the aggregated value of the received offers and the QoS end-to-end constraint. For one-issue negotiation in a composition of r services, such distance is calculated in a r-dimensional space. The same is for a multi-issue negotiation on r issues with one service. In both cases, the compositor accepts the offers when the distance is equal to 0.

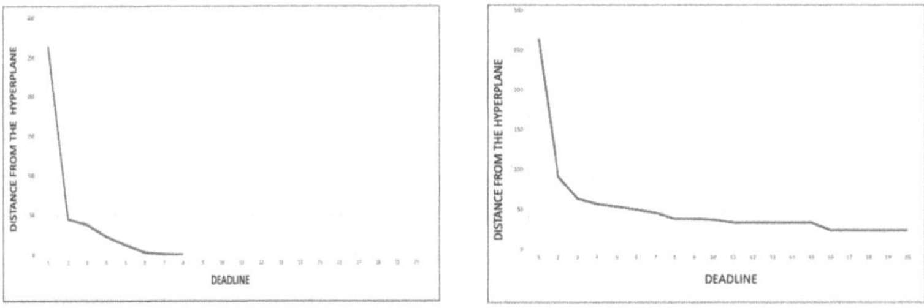

(a) Success case. **(b)** Failure case.

Fig. 3. Global distances trends in a case of successful and failed negotiations

4.1 Numerical Evaluation

We report a simple numerical simulation on the trends of the negotiation for the scenario reported in the previous section, considering 5 services in the SBA, and 4 provider agents for each service. In particular, we evaluate the trends of the offers received for each service by calculating, for all the offers, the utility of the offer provided by the jth provider for the ith service, with respect to the offers received by different providers for the same service (local evaluation), normalized with respect to the range of minimum and maximum values of the offers for all services. Such utility is computed, at each negotiation round, using the approach formulated in [11]:

$$U_{local}(price_{i,j}(t_k)) = \frac{max_i(price_{i,j}(t_k)) - price_{i,j}(t_k)}{\sum_{i=1}^{m} max_i(price_{i,j}(t_k)) - \sum_{i=1}^{m} min_i(price_{i,j}(t_k))}$$

where i identifies one of the m services (with $m = 5$) and the j identifies one of the n providers (with $n = 4$). Such utility is normalized to be in [0,1]. For each ith service the compositor agent selects the most promising offer ($price_{i,s}$), i.e. the one with maximum value of U_{local}. For each promising offer at round t_k, the global requirements satisfaction is evaluated by computing the Euclidean global distance of the composition of the values of the selected offers from the hyper-plane representing the end-to-end user's requirement (see Figures 3a and 3b). In the case of successful negotiations such distance (Figure 3a) converges to zero.

5 Conclusions

In the present work the use of software agent negotiation is used as a means to select service implementations required by an SBA by taking into account the Quality of Service that providers offer for their services, and the end-to-end QoS requirements expressed by a user requesting the application. We showed that negotiation in SBAs is inherently multi-dimensional even in the case of a single issue negotiation. Such multi-dimensions are given by the requirements of composing different services with provided QoS values that, once aggregated, meet an end-to-end constraint.

In this paper we showed that the negotiation implementation of a strategy on a single issue split among different provider agents available for the different services composing a requested SBA, can be easily extended for a multi-issue negotiation, so allowing to use the same negotiation mechanism, in terms of protocol and strategies, for both one- and multi-issue negotiation. Non linear utility functions, as well as concession strategies, can be modeled through the use of normal probability distributions, whose properties in scaling up in dimensions, allow to easily scale from one- to multi-issue negotiation by simply scaling the normal distribution dimensions.

References

1. Papazoglou, M., Traverso, P., Dustdar, S., Leymann, F.: Service-oriented computing: State of the art and research challenges. IEEE Computer 40(11), 38–45 (2007)
2. Yan, J., Kowalczyk, R., Lin, J., Chhetri, M.B., Goh, S.K., Zhang, J.: Autonomous service level agreement negotiation for service composition provision. Future Generation Computer Systems 23(6), 748–759 (2007)
3. Di Napoli, C., Pisa, P., Rossi, S.: Towards a dynamic negotiation mechanism for qoS-aware service markets. In: Pérez, J.B., Hermoso, R., Moreno, M.N., Rodríguez, J.M.C., Hirsch, B., Mathieu, P., Campbell, A., Suarez-Figueroa, M.C., Ortega, A., Adam, E., Navarro, E. (eds.) Trends in Pract. Appl. of Agents & Multiagent Syst. AISC, vol. 221, pp. 9–16. Springer, Heidelberg (2013)
4. Lau, R.Y.K.: Towards a web services and intelligent agents-based negotiation system for b2b ecommerce. Electronic Commerce Research and Applications 6(3), 260–273 (2007)
5. Lomuscio, A., Wooldridge, M., Jennings, N.: A classification scheme for negotiation in electronic commerce. Group Decision and Negotiation 12(1), 31–56 (2003)
6. Lai, G., Sycara, K.: A generic framework for automated multi-attribute negotiation. Group Decision and Negotiation 18(2), 169–187 (2009)
7. Gode, D.K., Sunder, S.: Allocative efficiency of markets with zero-intelligence traders: Market as a partial substitute for individual rationality. Journal of Political Economy 101(1), 119–137 (1993)
8. Baarslag, T., Hindriks, K.V.: Accepting optimally in automated negotiation with incomplete information. In: Proceedings of the 2013 International Conference on Autonomous Agents and Multi-agent Systems, AAMAS 2013, pp. 715–722 (2013)
9. Hanson, J.E., Tesauro, G., Kephart, J.O., Snible, E.C.: Multi-agent implementation of asymmetric protocol for bilateral negotiations (extended abstract). In: Proceedings of the 4th ACM Conference on Electronic Commerce, EC 2003, pp. 224–225. ACM, New York (2003)
10. Faratin, P., Sierra, C., Jennings, N.R.: Negotiation Decision Functions for Autonomous Agents. Robotics and Autonomous Systems 24, 3–4 (1998)
11. Alrifai, M., Risse, T.: Combining global optimization with local selection for efficient qos-aware service composition, pp. 881–890 (2009)

Strategic Bidding for Electricity Markets Negotiation Using Support Vector Machines*

Rafael Pereira[1], Tiago M. Sousa[1], Tiago Pinto[1], Isabel Praça[1], Zita Vale[1], and Hugo Morais[2]

[1] GECAD – Knowledge Engineering and Decision-Support Research Center, Institute of Engineering – Polytechnic of Porto (ISEP/IPP), Porto, Portugal
{1080730,tmsbs,tmcfp,icp,zav}@isep.ipp.pt
[2] Automation and Control Group – Technical University of Denmark, Denmark
morais@elektro.dtu.dk

Abstract. Energy systems worldwide are complex and challenging environments. Multi-agent based simulation platforms are increasing at a high rate, as they show to be a good option to study many issues related to these systems, as well as the involved players at act in this domain. In this scope the authors' research group has developed a multi-agent system: MASCEM (Multi-Agent System for Competitive Electricity Markets), which simulates the electricity markets environment. MASCEM is integrated with ALBidS (Adaptive Learning Strategic Bidding System) that works as a decision support system for market players. The ALBidS system allows MASCEM market negotiating players to take the best possible advantages from the market context. This paper presents the application of a Support Vector Machines (SVM) based approach to provide decision support to electricity market players. This strategy is tested and validated by being included in ALBidS and then compared with the application of an Artificial Neural Network, originating promising results. The proposed approach is tested and validated using real electricity markets data from MIBEL - Iberian market operator.

1 Introduction

The study of the operation of electricity markets has become increasingly important in recent years as a result of the challenges that the restructuring of these markets originated, this restructuring has increased the competitiveness of the market, but also its complexity. The increasing complexity and unpredictability, consequently, increase the difficulty in decision making. The bodies involved are then forced to rethink their behavior and market strategies [1], which makes it essential to use tools that allow the study and different market mechanisms and the relationships between the participating entities [2].

---

* This work is supported by FEDER Funds through COMPETE program and by National Funds through FCT under the projects FCOMP-01-0124-FEDER: PEst-OE/EEI/UI0760/2011, PTDC/SEN-ENR/122174/2010 and SFRH/BD/80632/2011 (Tiago Pinto PhD).

J.B. Perez et al. (eds.), *Trends in Practical Applications of Heterogeneous Multi-agent Systems. The PAAMS Collection*, Advances in Intelligent Systems and Computing 293,
DOI: 10.1007/978-3-319-07476-4_2, © Springer International Publishing Switzerland 2014

Regulators and players have an interest in predicting market behavior; regulators to test regulations before they are implemented, and detect market inefficiencies, the market players to realize market behavior in order to optimize their profits.

The need to understand these mechanisms and how the interactions between the players affect the market has contributed to the increased use of simulation tools. Multi-agent software is particularly directed to the analysis of dynamic and adaptive systems with complex interactions among stakeholders. Several modeling tools for the study of electricity markets have emerged. Some relevant examples are Electricity Market Complex Adaptive System (EMCAS) [3], Agent-based Modelling of Electricity Systems (AMES) [4], Genoa Artificial Power Exchange (GAPEX) [5], and Multi-Agent System for Competitive Electricity Markets (MASCEM) [6, 7].

Although some studies have emerged, confirming the applicability of simulation tools to study these markets, particularly using multi-agent systems, these tools present a common limitation: the lack of adaptive learning capabilities that enable them to provide effective support to the decisions of market entities. Current tools are directed to the study of market mechanisms and interactions among participants, but are not suitable for supporting the decision of the players' negotiators in obtaining higher profits in energy transactions.

These limitations highlight the need to develop adaptive tools, which enable strong support for market players. These tools will enhance the improvement of the results of these players. Being adaptable to different market circumstances and contexts of negotiation, intelligent tools can get realistic and appropriate suggestions for players' actions so that they can direct their behavior in search of the best possible results.

With a view to eliminating this gap, the Adaptive Learning strategic Bidding System (ALBidS) [7, 8] was developed. ALBidS is based on multi-agent technology, assembling several different strategic approaches to act in the electricity market.

The ALBidS system is integrated with the MASCEM simulator, a system that models the environment of electricity market and the interactions between the major participating organizations, both in negotiation and management.

The competitive nature of electricity markets, which translates to constant and rapid changes in this environment, also require an endless search for new methods of artificial intelligence, adaptive learning and decision support, enabling systems like ALBidS adequate adaptation at all times. The nature of these new approaches should reflect an increasing efficacy in forecasts in the shortest time possible execution.

Support Vector Machines (SVM) [9] is a technique for classification and data forecasting (closely related to artificial neural networks (ANN) [10]), which, by its characteristics, offers some guarantees to grant ALBidS with the required specifications: good predictive power in short execution time.

SVMs have been used in many areas, such as pattern recognition, image recognition, classification and regression analysis, text categorization, medical science to classify proteins, weather forecast, energy prices, and many other applications [11, 12].

This paper presents the development of an approach based on SVM, so that it can be integrated in the ALBidS system [8], in order to make predictions of electricity market prices, in order to be used as a basis for operational strategy of agents participating in the electricity market.

The interconnection with ALBidS and MASCEM [6] allows the testing and validation of the proposed strategy in a realistic electricity market simulation environment, using real data from the Iberian power market - MIBEL [13].

1 MASCEM and ALBidS Overview

MASCEM [6, 7] is a multi-agent system that models the most important players that take part in electricity markets, collecting data in the medium and long term to support the decisions of these agents according to their characteristics and objectives, thus allowing better understanding of the mechanisms and behaviors that are common to this type of markets. The MASCEM simulator uses several approaches and learning techniques for modeling and supporting market actors in their decisions. Fig. 1 shows the main features of the MASCEM simulator.

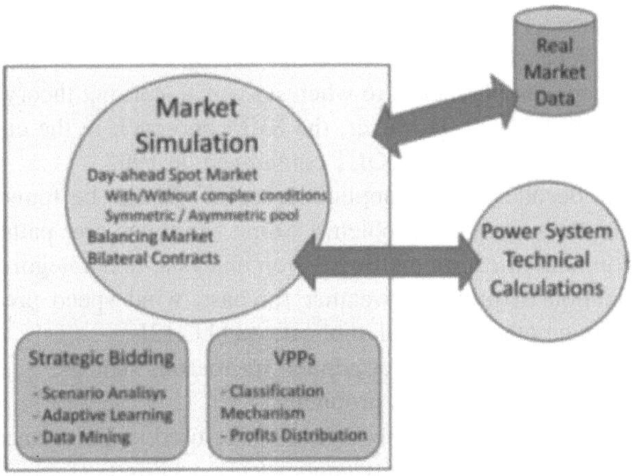

Fig. 1. MASCEM key features [14]

The simulator is directed to the study of various types of markets, including the day-ahead spot market, balancing markets, forwards markets, and bilateral contracts. MASCEM allows defining the simulation settings and market scenarios to simulate, including the definition of the characteristics of buyers, sellers, Virtual Players (VPP) and their number. These scenarios are supported by real data from several electricity market operators [13], which provides a realistic representation of such markets.

After the simulations it is possible to analyze the amounts of energy transacted by each agent, the respective market prices, and the obtained profits/costs in each period.

ALBidS (Adaptive Learning strategic Bidding System) [7, 8] is a multi-agent system, which integrated with MASCEM. ALBidS uses adaptive learning to make MASCEM agents capable of analyzing negotiation contexts, such as the types of markets in which they are inserted, economic and weather conditions, the type of day (*e.g.* business, weekend, or holiday) and the trading period (*e.g.* peak or off-peak).

Thus, agents can adapt to any market automatically, being prepared for unexpected changes in electricity markets, managing to change their strategy of buying or selling through this system. For this, ALBidS uses Reinforcement Learning Algorithms (RLA) to choose the strategy that best fits each specific context.

The Main Agent, which is the top entity in ALBidS, is responsible for running the RLAs; allowing ALBidS to choose the best strategy that most suits each specific situation. The strategies are all independent from each other, and based on different approaches and techniques, being executed in parallel by a different agent.

2 Strategic Bidding Based on Support Vector Machines

In 1936, R. A. Fisher [9] created the first algorithm for pattern recognition. Subsequently, Vapnik and Lerner in [15] created an algorithm called Generalized Portrait (SVM algorithm is implemented by a generalization of the nonlinear algorithm Generalized Portrait). This was the first running kernel of SVM, only for classification and linear problems.

The SVM concept can be tracked to when statistical learning theory was developed further with Vapnik, in 1979. However, the SVM approach in the current form was first introduced with a paper at the COLT conference, in 1992.

During the last decades several applications of SVM can be found [16], both for classification and for regression problems. Some examples are: pattern recognition, image recognition, classification and regression analysis, text categorization, medical science, classification of proteins, weather forecast, wind speed prediction, energy prices forecast, among other practical applications [11, 12].

SVM application using *Kernels* got popular for reasons such as:

- Often concentrating on convex problems;
- Allowing many linear algebra techniques to be used in a non-linear way;
- Have showed robustness in many application domains;
- Spend fewer resources and half the time of artificial neural networks.

The information to use in an SVM must follow the format suggested in (1):

$$(y_1, \mathbf{x}_1), \ldots, (y_l, \mathbf{x}_l), x \in R^n, y \in R, \tag{1}$$

where each example x_i is a space vector example; y_i has a corresponding value; n is the size of training data. For classification: y_i assumes finite values; in binary classifications: $y_i \in \{+1,-1\}$; in digit recognition : $y_i \in \{1,2,3,4,5,6,7,8,9,0\}$; and for regression purposes, y_i is a real number ($y_i \in$ R).

The implementation of SVM requires considering some important aspects, namely:

- **Feature Space** is the method that can be used to construct a mapping into a high dimensional feature space by the use of reproducing *kernels*. The idea of the *kernel* function is to enable operations to be performed in the input space rather than the potentially high dimensional feature space. Hence the inner product does not need to be evaluated in the feature space. This provides a way of addressing the curse of dimensionality. However, the computation is still highly dependent on the number of training patterns, and a good data distribution for a high dimensional problem generally requires large training sets.

- **Loss Functions**. In statistics, the decision theory and machine learning, the loss function is a function that maps an event to a real number, representing some "costs" associated with the difference between the estimated and the actual data for an occasion. The purpose of this function is to modulate the input data, when applied to a training set, and then forecasting the values (or sorting). The loss function uses the forecast values and compares how much they deviate from the actual values, quantifying the deviation.
- ***Kernel* Functions**. The *kernel* functions, in general, are a set of algorithms for pattern examination. The main task is to find patterns and study the type of associations, in a particular pattern (*e.g*, groups, classifications, major components, correlations, classifications) for general types of data (such as sequences, text documents, sets of points vectors, images, etc). The kernel function approach the problem by mapping the data to a dimensional space, where each coordinate corresponds to a characteristic of each input value transforming the data into a set of points in Euclidean space. Some examples of kernels are [17]: Polynomial, Gaussian Radial Basis Function, Exponential Radial Basis Function, Multi-Layer Perceptron, Splines, B splines.

The most applicable *kernels* for the specific problem approached in this paper (forecast of electricity market prices) are the Radial Basis Function (RBF) and the Exponential Radial Basis Function (eRBF). These two *kernels* are directed to regression in time series data.

The development of the SVM approach was performed in MATLAB, which is a framework highly directed to mathematical calculations, such as required by this problem. The SVM approach for regression of the electricity market prices, takes as parameters:

- *trainingLimit* - limit number of training days;
- *kernel* - kernel that will be used in the regression process;
- *e_Val* - Value of ε-insensitive;
- *C* - limit;
- *p1* – angle (σ).
- *p2* - offset.

In order to run the SVM, given the date and period of the day for which the SVM will predict the market price, the application automatically creates a MS Excel file with all the data necessary for training, which is gathered from the MASCEM database of historic market prices. The file contains a worksheet with the name *Training*, which contains a column with the number of lines that were set as the limit for training. Then, the SVM approach is ran with the specified parameters, and the forecasted price is returned to the application to be used by ALBidS.

3 Case Study

This section demonstrates some results of the tests that were performed to the implemented SVM approach. The SVM performance is compared to a previous implementation of an artificial neural network (ANN), which is already integrated in ALBidS, with the same purpose as the proposed SVM approach – forecasting electricity market prices to provide decision support to electricity market negotiating players [8]. The comparison of results is performed in three different dimensions: (i) comparison of the forecasting error using the Mean Absolute Percentage Error (MAPE); (ii) comparison of the execution times when executing the SVM and the ANN; (iii) comparison of the incomes originated to an electricity market player, when used as decision support in market negotiations. The electricity market negotiations are simulated in MASCEM, using a scenario based or real data from the Iberian market operator – MIBEL [13]. The simulations, as well as all the tests concern 61 consecutive days (2 months), starting on September 1st, 2009.

All results are analyzed by period independently, since in the scope of ALBidS, periods are considered as different, independent contexts [8].

Regarding the SVM parameterization, two *kernels* are considered: RBF and eRBF.

After exhaustive sensibility analysis, the parameterizations that achieved the best results, and the ones used in this case study, are:

- *kernel* RBF: *Training Limit* = 20, $p1(\sigma) = 6$, ε-insensitive = 0, $C = \infty$, $p2 = 0$;
- *kernel* eRBF: *Training Limit* = 20, $p1(\sigma)=18$, ε-insensitive = 0, $C = \infty$, $p2 = 0$

Table 1 presents the average MAPE values of forecast of the SVM approach using each of two *kernels* (RBF, and eRBF) and the ANN, for the 61 considered days.

Table 1. MAPE forecast error values (%)

Period	SVM (RBF)	SVM (eRBF)	ANN
1	0,297578	0,262867	0,326704
6	0,044731	0,107784	0,02703
12	0,788205	0,385893	0,579867
18	0,033467	0,099395	0,003859
Total	1,163981	0,855939	0,93746

From Table 1 it is visible that the error values are always located below 1%. Comparing the performance of the SVM with the ANN one can see that the forecast results are very similar for all cases. In the total of all error values for the four considered periods of the 61 days, the SVM approach using the RBF *kernel* achieved a higher error than the ANN, while the SVM approach using the eRBF *kernel* was able to achieve lower error values than the ANN. However, the error values are so similar, that conclusions on what is the best approach for forecasting electricity market prices cannot be taken. Table 2 presents the average execution times of the SVM approach with different amounts of training data, after 1000 run trials.

Table 2. Average execution times of the SVM approach, in milliseconds

	Execution times (in m.s.)		
Training Limit	**Minimum**	**Average**	**Maximum**
5	4776	5022	9501
10	4730	4984	5830
15	4721	5107	7437
20	4789	5092	10136
35	4793	4942	23595

Table 2 shows that the SVM approach, using either of the two considered *kernels*, takes an average of 5 seconds to execute. From Table 2 it is also visible that the increase in considered days for training does not represent a considerable degradation in execution time. Table 3 presents the average execution times of the ANN approach, for different amounts of training data.

Table 3. Average execution times of the ANN approach, in milliseconds

	Execution Time (milliseconds)	
Training Limit (days)	**Without Parallelism**	**With Parallelism**
60	15 000	11 000
120	18 000	13 000
200	22 000	14 000
365	30 000	17 000
730	49 000	20 000

Table 3 shows the average execution time of the ANN. It is visible that, even with parallel programming for a faster access to data, the minimum average value is of 11 seconds (more than twice the average execution time of the SVM). Note that the amount of data that the ANN requires for its training is enormous when comparing to the SVM (ANN: 60 to 730 days to achieve acceptable results; SVM: 5 to 35 days).

Finally, Table 4 presents the comparison of the incomes that were achieved in the simulated electricity market, by a player using the SVM and the ANN as decision support for market negotiations.

Table 4. Incomes achieved in the simulated electricity market, using the both *kernels* of the SVM and the ANN, in €

Negotiation Period	SVM (RBF)	SVM (eRBF)	ANN
1	7391,65	7874,5	5121,21
6	13187,08	16069,33	4190,76
12	9551,25	9036,00	6095,14
18	18157,39	22398,50	16511,02
Total	48287,37	55378,33	31918,13

From Table 4 it is visible that both the SVM, using each *kernel,* achieved higher incomes than the ANN approach. Given the similarity in the SVM and ANN forecasting capabilities for this problem, as shown in Table 1, these results are explained by the fact that the SVM forecasting errors are often by defect (below the real value), which originates that the player is always able to sell, even if at a bit smaller price; while when bidding above the market price (as happens more often with the ANN) means not selling at all.

4 Conclusions

This paper presented a methodology for electricity market negotiation, based on the concept of Support Vector Machine. SVMs are widely used in the last decades, and can be applied to classification and regression problems. The main advantages are the good forecasting capabilities, and the small amount data required for the training process, which leads to very low execution times.

The proposed approach was integrated with ALBidS, a decision support system for electricity market players, which is integrated with the electricity market simulator MASCEM. This integration provided the means for the testing of the proposed approach under a realistic simulation scenario based on the Iberian electricity market – MIBEL. The results achieved with the SVM approach are encouraging, in a sense that, for both considered *kernels*, the forecasting achieved very good results (always below 1% error), and very similar to the performance of the ANN. However, in what concerns the execution time, the SVM approach showed all of its advantage, by taking less than half the time of the ANN to provide results, by using only the data of a few days for training, while the ANN requires several months. When supporting a market player's decisions the electricity market, the SVM approach was able to provide higher incomes than the ANN.

References

1. Shahidehpour, M., et al.: Market Operations in Electric Power Systems: Forecasting, Scheduling, and Risk Management, pp. 233–274. Wiley-IEEE Press (2002)
2. Meeus, L., et al.: Development of the Internal Electricity Market in Europe. The Electricity Journal 18(6), 25–35 (2005)
3. Koritarov, V.: Real-World Market Representation with Agents: Modeling the Electricity Market as a Complex Adaptive System with an Agent-Based Approach. IEEE Power & Energy magazine, 39–46 (2004)
4. Li, H., Tesfatsion, L.: Development of Open Source Software for Power Market Research: The AMES Test Bed. Journal of Energy Markets 2(2), 111–128 (2009)
5. Cincotti, S., Gallo, G.: Genoa Artificial Power-Exchange. Agents Artif. Intell. (2013)
6. Praça, I., et al.: MASCEM: A Multi-Agent System that Simulates Competitive Electricity Markets. IEEE Intelligent Systems 18(6), 54–60 (2003)
7. Vale, Z., et al.: MASCEM - Electricity markets simulation with strategically acting players. IEEE Intelligent Systems 26(2) (2011); Special Issue on AI in Power Systems and Energy Markets

8. Pinto, T., Vale, Z., Rodrigues, F., Morais, H., Praça, I.: Strategic Bidding Methodology for Electricity Markets Using Adaptive Learning. In: Mehrotra, K.G., Mohan, C.K., Oh, J.C., Varshney, P.K., Ali, M. (eds.) IEA/AIE 2011, Part II. LNCS, vol. 6704, pp. 490–500. Springer, Heidelberg (2011)
9. Fisher, R.: The use of multiple measurements in taxonomic problems. Annals of Eugenics 7, 111–132 (1936)
10. Amjady, N., et al.: Day-ahead electricity price forecasting by modified relief algorithm and hybrid neural network. IET Generation, Transmission & Distribution 4(3), 432–444 (2010)
11. Cortez, P., et al.: " Modeling wine preferences by data mining from physicochemical properties", Department of Information Systems/R&D Centre Algoritmi, University of Minho, 4800-058 Guimarães, Portugal (2009)
12. Mohandes, M., et al.: Support vector machines for wind speed prediction., Department of Electrical Engineering, King Fahd University of Petroleum and Minerals, Dhahran 31261, Saudi Arabia (2004)
13. MIBEL – Operador del Mercado Iberico de Energia website, (acessed on January 2013), `http://www.omie.es/`
14. Pinto, T., et al.: A new approach for multi-agent coalition formation and management in the scope of electricity markets. Energy Journal (2011), doi:10.1016/j.energy.2011.05.045
15. Vapnik, V., Lerner, A.: Pattern recognition using generalized portrait method. Automation and Remote Control 24, 774–780 (1963)
16. Vapnik, V., et al.: The Nature of Statistical Learning Theory. Springer-Verlag New York, Inc. (1995)
17. Gunn, S.: Support Vector Machines for Classification and Regression. Technical Report, Image Speech and Intelligent Systems Research Group, University of Southampton (1998)

A Mobile Robot Agent for Gas Leak Source Detection

Dani Martínez[1], Tomàs Pallejà[1], Javier Moreno[1], Marcel Tresanchez[1],
Mercè Teixidó[1], Davinia Font[1], Antonio Pardo[2], Santiago Marco[2,3], and Jordi Palacín[1]

[1] Department of Computer Science and Industrial Engineering,
University of Lleida, 25001 Lleida, Spain
{dmartinez,tpalleja,jmoreno,mtresanchez,
mteixido,dfont,palacin}@diei.udl.cat
[2] Department of Electronics, University of Barcelona, Martí i Franquès, 1,
08028 Barcelona, Spain
apardo@el.ub.es
[3] Signal and Information Processing for Sensing Systems,
Institute for BioEngineering of Catalonia, Baldiri Reixac, 10-12, 08028 Barcelona, Spain
smarco@ibecbarcelona.eu

Abstract. This paper presents an autonomous agent for gas leak source detection. The main objective of the robot is to estimate the localization of the gas leak source in an indoor environment without any human intervention. The agent implements an SLAM procedure to scan and map the indoor area. The mobile robot samples gas concentrations with a gas and a wind sensor in order to estimate the source of the gas leak. The mobile robot agent will use the information obtained from the onboard sensors in order to define an efficient scanning path. This paper describes the measurement results obtained in a long corridor with a gas leak source placed close to a wall.

Keywords: gas detection, mobile robot agent, laser sensor, self-localization.

1 Introduction

There are some accidental and harmful situations such as gas leak accidents in which humans expose their health to risky factors and situations. Currently, computers and sensor technological advances have allowed the development teleoperated robots [1] and artificial agents in order to substitute humans in risky tasks. In [2] was proposed an agent architecture for risk perception and interpretation taking into account different dimensions. Furthermore, a multi-agent system for ambient assistance based on user behavior learning was implemented and assessed [3].

On the other hand, gas leakage localization is a well-known problem in chemical robotic applications. Electronic noses are the most used devices for gas detection in many applications; nevertheless, there is a high difficulty in source leak location using these sensors due to the chaotic diffusion of the gas in the air [4]. In the bibliography there are several papers where uses autonomous mobile robot agents equipped with odor sensors as input for specific algorithms to predict the localization of gas leak

J.B. Perez et al. (eds.), *Trends in Practical Applications of Heterogeneous Multi-agent Systems. The PAAMS Collection*, Advances in Intelligent Systems and Computing 293,
DOI: 10.1007/978-3-319-07476-4_3,

sources [5, 6]. In robotics and in other autonomous systems often uses a laser sensors to detect and locate objects or obstacles with high fidelity. In [7] is described an object localization procedure using a static Light Detection and Ranging (LIDAR) sensor for mobile robot tracking and discusses the location errors found.

This paper simulates a gas leak accident in an indoor scenario in which the air is mixed with Acetone, always with low concentrations in order to keep the air breathable for humans without having any risk during the experimentation. Figure 1 presents the robot agent developed by the research group which is capable to auto-localize itself in an indoor environment capturing data with its LIDAR sensor and perform a Simultaneous Localization and Mapping procedure (SLAM). The final milestone of this work is to implement an effective robot agent which will be able to estimate the localization of the source gas leak scanning and processing the data taken from the photo ionization detector (PID) and the other sensors for each position. This paper reflects closely the processes performed by the robot agent for reaching its functional objective.

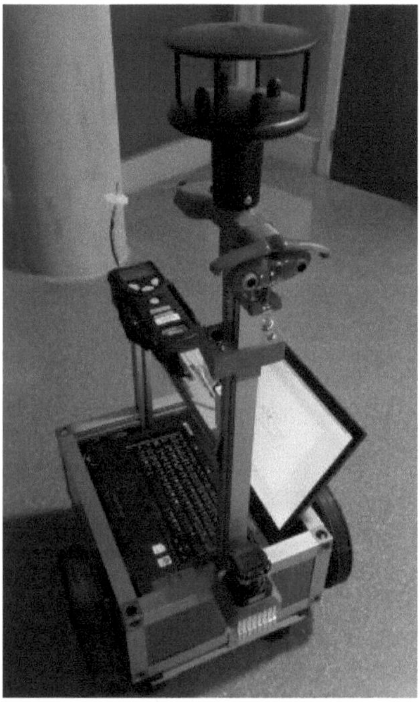

Fig. 1. Mobile robot agent developed, rBot

2 Materials and Methods

The mobile robot agent developed has the following sensors attached: a LIDAR, an anemometer, and a photo ionization detector (PID). For the experimentation is also used a gas source mechanism to simulate the gas leakage. Additionally, the procedure for simultaneous location and mapping (SLAM) is considered as a method in this work.

2.1 LIDAR Sensor

The robot agent uses a Hokuyo UTM-30LX Laser (Figure 2) for mapping, self-localization and obstacle avoidance. This sensor scans the distance between the sensor and the obstacle, with a range from 0.1 to 30 m. However, in this paper, the maximum range has been limited to 4 m. Each sensor scan takes approximately 25 ms and gathers 1.081 distance points between the angles -135° to +135° where the 0° is the center of the sensor. Lectures are represented in polar coordinates and distance component has a resolution of 1mm. It is placed in front of the robot and the computed samples are sent using an USB 2.0 interface.

Fig. 2. Hokuyo UTM-30LX LIDAR sensor

2.2 Anemometer

The robot agent is equipped with an anemometer on the top of its build for detecting the wind speed and its direction. The WindSonic model by Gill Instruments Ltd. obtains the velocity of the wind in m/s and the direction angles in degrees, both lectures at 4Hz of maximum sampler frequency. The communication with this device is performed by using a USB simulating a RS232 interface. The information obtained by this sensor will be used as an input parameter for the estimation algorithms of gas source leak localization.

2.3 Photo Ionization Detector (PID)

A PID ppbRAE 3000 sensor by RAE Systems is installed in the robot to detect the gas concentration in the air. This sensor can detect the concentration of different types of gases, so the device must be configured to detect acetone gas. Samples are obtained in parts per billion (ppb) with 2 seconds of response time and are sent by RS232 interface, but it can also be communicated through Bluetooth or USB interfaces. As the previous device, sample values are also used for the gas source leak localization algorithms as input parameter.

2.4 Mobile Robot Agent

A Hewlett-Packard laptop Intel Core2 Duo @1.66GHz, 1GB of RAM, is used as the core of the mobile robot and runs with Windows XP Operating System. All the previous sensors are connected to this computer directly. In addition, a motor control board is installed and connected with the laptop through USB connection. To engage the communication, the system creates a virtual serial port, so it can be used as a standard serial interface from a terminal or any command sending script. Using the motor encoders, the robot agent can perform discrete movements. For example: go straight 1m, or turn left 40º. The laptop and the other robot devices are powered with two external batteries of 95 W/h. The control program initializes all the communication interfaces used in the mobile robot and then executes the proposed mobile robot agent implementation.

2.5 Gas Leak Source

A mechanism has been built in order to simulate a gas leak source (Figure 3). In this case some Acetone is placed in the bottom of the vertical plastic pipe which has a fan on the top. So, this mechanism extracts the air out and evaporates the acetone at 32 ml/h in average.

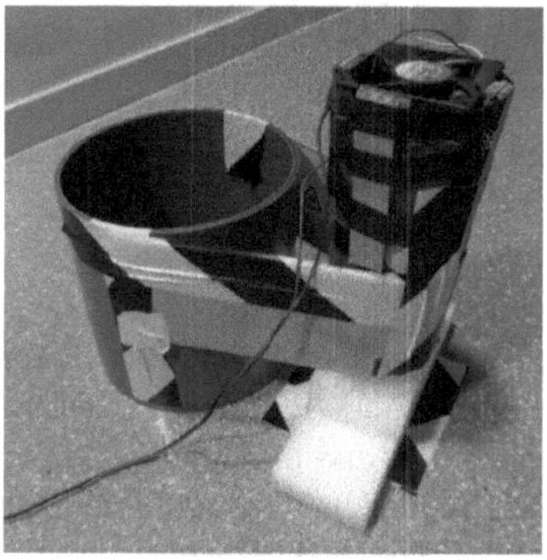

Fig. 3. Acetone evaporator mechanism

2.6 SLAM

The robot self-localization procedure requires an SLAM methodology to estimate the displacement offset in real-time. The odometry [8] obtained from the encoder of the motion motors is also available as a comparative reference. The SLAM method consists on comparing laser scans in real-time. This comparison is based on the template matching method between the reference data obtained with the LIDAR (explored map) and the new acquired data by the laser. The comparison returns a position offset (Δx, Δy) and a relative rotation angle (φ), so, the map would be built while the mobile robot is exploring the area. The explored map is represented as a two-dimensional boolean matrix.

3 Agent Methodology

The proposal of this paper is the development of a fully autonomous agent that will obtain data from the environment in order to develop the planned tasks. The agent will also decide if the final objective has been completed and provide the localization of the gas leak source). Figure 4 shows different steps included in the agent methodology.

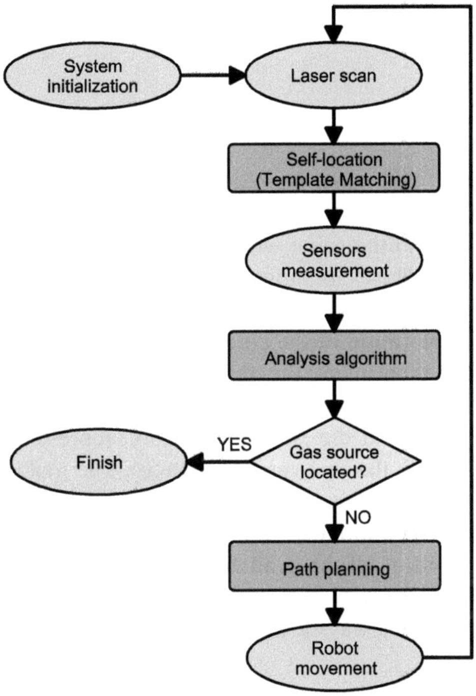

Fig. 4. Robot Agent behavior diagram

In the first step, the robot initializes and checks all the devices and sensors connected physically to it. At this point, the agent performs the first laser scan and stores it as the initial reference map, thus, there is no self-location process executed. Then, the agent tries to detect reference elements such as straight walls from the initial reference map in order to guide the initial displacement mobile robot and initiate a complete exploration of the area in front of the mobile robot. The mobile robot agent proposes an exploratory path planning depending on the information extracted on the reference map. For instance, in small spaces fully covered in the initial reference map (the LIDAR scan provides information of up to 40 m) the robot will perform a right wall exploration until arriving to the starting position. Otherwise, in larger areas, the robot can perform a wall-to-wall zig-zag exploration and it can be configured to perform random path planning exploration in order to compare its effectiveness against the zig-zag displacement. The mobile robot also used the LIDAR information to avoid unexpected obstacles.

In the next iterations the mobile robot agent obtains scan data from the LIDAR and performs the self-location procedure. The new gathered data is compared with the reference map and the displacement is processed in a heavy template matching procedure implemented in C-code for speed optimization. Simultaneously, the agent completes gas, wind and odometry measurement. The information obtained with the sensors is used to update a localization map that is also used to compute the estimate of the gas leak source. In this paper the hypothesis is that the gas source position estimate will be more accurate as the information of the exploration is more complete.

4 Results

The experimental part of this paper was carried out at the second floor of the Polytechnic School of the University of Lleida. This area has a large and tight corridor of approximately 40m long, so the robot will have the opportunity to explore it by following its right-side wall. The evaporation mechanism was placed near the wall of the corridor at a centric localization to simulate a gas leak accident. The mobile robot performs a complete exploration of the corridor by measuring gas concentration and relative position. In this experiment, the gas leak source is estimated as the center of mass (centroid) of such information. Figure 5 shows the map of the corridor obtained with the SLAM procedure and the gas concentration represented as colored circles (the yellow circle represents the true gas leak source). The mobile robot agent implemented samples gas concentration at approximately 1 Hz with a relative maximum displacement of 0.6 m/s. The mobile robot agent updates a gas leak source estimate by computing the centroid of all the available gas concentrations until detecting that the unknown area was completely explored. The error obtained in the final gas leak source position estimate was always lower than 1 m.

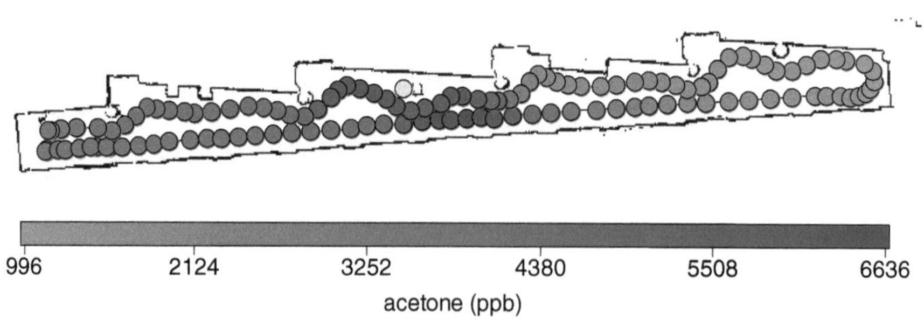

Fig. 5. Corridor map obtained with the SLAM procedure, relative position of the gas concentrations (colored circles) and position of the gas leak source (yellow circle)

5 Conclusion

This paper proposes the implementation of a mobile robot agent for gas leak source detection in human-risky indoor conditions. The mobile robot carries several sensors in order to analyze and recognize the environment and measure gas concentration in the air. The objective of the agent is estimate the location of a unique gas leak source. The mobile robot agent is designed to explore an unknown area by using a SLAM procedure while sampling gas concentration and air displacement in order to provide a gas leak source position estimate performed by computing the centroid of the gas concentration values. Results showed that the proposed mobile robot agent estimate the gas leak source position with an approximated error of 1m. Future works will be focused on improving the gas detection procedure by mixing the information of the displacement of the air provided with the onboard wind sensor and the gas concentration measured [9].

Acknowledgements. This work was partially funded by the Spanish Ministery of Economy and Competitivity, Plan Nacional de Investigación Científica, Desarrollo e Innovación Tecnológica: TEC2011-26143, and by the Government of Catalonia (Comisionat per a Universitats i Recerca, Departament d'Innovació, Universitats i Empresa) and the European Social Fund.

References

1. Woosub, L., Sungchul, K., Munsang, K., Mignon, P.: ROBHAZ-DT3: teleoperated mobile platform with passively adaptive double-track for hazardous environment applications. In: 2004 IEEE/RSJ International Conference on Intelligent Robots and Systems (IROS), Sendai, Japan, vol. 1, pp. 33–38 (2004)
2. Trindade, N., Antunes, L.: An Architecture for Agent's Risk Perception. Advances in Distributed Computing and Artificial Intelligence Journal 1(5), 75–85 (2013)
3. Guivarch, V., Camps, V., Péninou, A.: AMADEUS: an adaptive multi-agent system to learn a user's recurring actions in ambient systems. Advances in Distributed Computing and Artificial Intelligence Journal 1(3), 1–10 (2012)
4. Shraiman, B.I., Siggla, E.D.: Scalar turbulence. Nature 405(6787), 639–646 (2000)
5. Hiroshi, I., Takamichi, N., Toyosaka, M.: Remote sensing of gas/odor source location and concentration distribution using mobile system. Sensors and Actuators B: Chemical 49(1-2), 52–57 (1998)
6. Ishida, H., Tanaka, H., Taniguchi, H., Moriizumi, T.: Mobile robot navigation using vision and olfaction to search for a gas/odor source. Autonomous Robots 20(3), 231–238 (2006)
7. Teixidó, M., Pallejà, T., Font, D., Tresanchez, M., Moreno, J., Palacín, J.: Two-Dimensional Radial Laser Scanning for Circular Marker Detection and External Mobile Robot Tracking. Sensors 12, 16482–16497 (2012)
8. Borenstein, J., Liqiang, F.: Measurement and correction of systematic odometry errors in mobile robots. IEEE Transactions on Robotics and Automation 12(6), 869–880 (1996)
9. Pomareda, V., Hernández, V., Khaliq, A.A., Trincavelli, M., Lilienthal, A.J., Marco, S.: Chemical source localization in real environments integrating chemical concentrations in a probabilistic plume mapping approach. In: 15th International Symposium on Olfaction and Electronic Nose, ISOEN (2013)

RETRACTED CHAPTER: Integration of Different ERP Systems on Mobile Devices

Álvaro Lozano[1], Ana Belen Gil[1], and Tiancheng Li[2]

[1] Department of Computer Science and Automation, University of Salamanca
Plaza de la Merced, s/n, 37008, Salamanca, Spain
{loza,abg}@usal.es

[2] School of Mechatronics, Northwestern Polytechnical University,
Xi'an, 710072, P.R. China
t.c.li@mail.nwpu.edu.cn

Abstract. Nowadays a lot of enterprises work with ERP systems. It uses times, is generally used in office environments and different enterprises which offer this software are developing mobile applications. These mobile applications work with their own system and they don't usually work in other platforms. Currently any mobile application can communicate with more than one ERP system because each one has its own communications methods. This article presents a system that expect unify the communication between different ERP systems and allows mobile applications to communicate with them in a homogeneous way.

Keywords: ERP, Mobile, VTD-XML, XPATH, REST, Android.

1 Introduction

Nowadays a lot of enterprises decide to start using ERP (Enterprise Resource Planning). These systems manage a lot of daily tasks of the different enterprises sectors: production, logistics, allocation, stock, consignment, invoicing, etc. Their implementations can be adapted to the business that they manage [1]. Currently ERP systems can be found in office environments and managed by qualified staff. This kind of software requires a previous training to take advantage of it.

The mobiles devices are every time smaller, lighter and they have improved the performance, these are the reason why it seems a good idea transferring part of the ERP usefulness to these devices [3].

Transferring ERP feasible process to a mobile device will allow an enterprise staff outside of the office to make distance specific operations on the ERP software thanks to nice interface normally presented on mobile applications.

These applications aim to transform mobile devices into work tools that are completely integrated into staff's daily work, getting closer to typical notions of pervasive computing [6] Enterprises not only will use the applications in these mobile devices as work tools, but also to gather information and thus make better their internal processes based on acquired knowledge achieved by that information [7]

The original version of this chapter was retracted: The retraction note to this chapter is available at https://doi.org/10.1007/978-3-319-07476-4_22

J.B. Perez et al. (eds.), *Trends in Practical Applications of Heterogeneous Multi-agent, Systems. The PAAMS Collection*, Advances in Intelligent Systems and Computing 293,
DOI: 10.1007/978-3-319-07476-4_4, © Springer International Publishing
Switzerland 2014, corrected publication 2024

Nowadays several companies that commercialize ERP systems are directing this software to SaaS [2] [28] [31] [30] Architectures and they also have platforms to develop applications EMA (Enterprise Mobile Application) [4] [26] [32] which make possible linking mobile devices of different platforms and the ERP software offered by them.

The main trouble is the fact that different ERPs don't have the same public interface, that is, we don't have neither a communication protocol nor common information [8] [9]. Currently the enterprises who offer ERP software invest a lot of time and money in the development of native applications on different platforms to use them with their own ERP. This means that there are no applications which can link with other ERPs.

Along this paper a system to integrate different ERPs is presented as a possible solution of this trouble. We expect to unify different linking methods of the different ERP systems by developing a standard of communication and information and adapting this information of every system to the standard. This will allow enterprises to develop mobile applications which are able to operate with different ERP systems.

This article is divided as follows: section 2 describes the state of the art; section 3 presents the proposed model; section four describes the results obtained and the conclusions respectively.

2 Background

There are many applications that ERP Software companies offer to customers. Such as SAP Business One which has management applications on Windows Phone, IOS and Android platforms. Exist both web applications as native applications on the market but they all have operations in the ERP for which they were designed [37] [38] [39]. Companies usually distribute this kind of software they certify official partners who are responsible for developing applications. Official Partners use SDKs, APIs or communication offered by ERP systems with the outside which are commonly SOAP web services [5] [2[illegible]] [25] [27] [29].

Due to communication with the outside of the ERP systems, companies as Jitterbit [34] have emerged. Jitterbit performs transformations of data streams of different types of web services.

This paper aims to transform not only the data but set a standard to unify communication and information that can be given between these ERP systems and mobile or web applications.

The ERP market is varied and there are many different systems from diverse companies. The best known is chosen SAP and despite only having a market share of 22% [33] [10] [21] [23] [24].

3 Proposed Reasoning System

The proposed system aims to create an integration of different ERPs so that it provides a common interface to all of them. It facilitates the creation of mobile applications that are compatible with multiple ERP systems. The system initially presents the integration of three ERP systems from two leading companies in the sector [10] such as Microsoft and SAP. ERP systems are initially chosen Microsoft Dynamics NAV, Microsoft Dynamics AX and SAP Business One.

The system will display the following general structure figure 1:

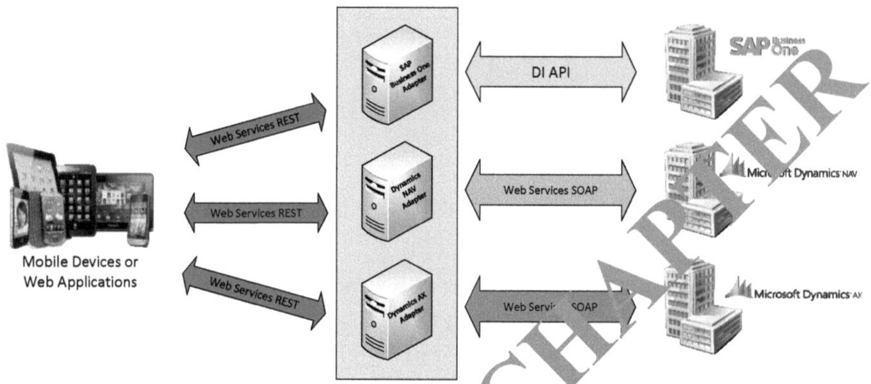

Fig. 1. General structure

For each one of the ERPs integrated in the system there should be an adapter where the ERP information will be transformed to the established information standard and vice versa.

The kind of communication that will exist between mobile devices and adapters will be homogeneous. This communication is based on a set of REST services [20] [19] [18] that will may be used by web or mobile applications.

The development of this system poses two challenges mainly:

- Perform a standard of information between ERP systems integrated. This must be scalable in order to integrate new ERPs in the system.
- Establish a communication system scalable, fast and homogeneous for communication with mobile devices.

To create information standard, a thorough analysis of the information which is handled by different ERP, should be made. Several information entities covering different modules that manage ERP systems will be created. The information for each entity will be different in each ERP system so it will be necessary to develop correlation between ERP information and information pertaining to the information standard. Subsequently established correspondences guide the transformation of the information carried out on each adapter.

The conversion information is implemented by performing transformations of XML files (Figure 2). For processing and more efficient parsing VTD-XML [36] will be used.

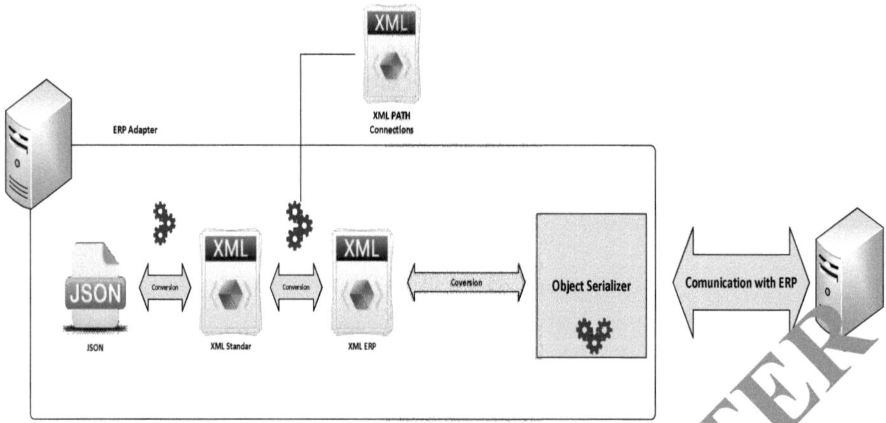

Fig. 2. Conversion of information in ERP Adapter

The communication interface that exist between mobile devices and adapters will be a collection of REST services common to all adapters. Web or mobile applications will communicate with REST services [35] [20] [19] [18] to perform operations on the ERP system. Each adapter must establish what services are active depending on the possibilities offered by the ERP system in each case.

4 Results

The mobilization of information and pre-sales and route accounting operations of the ERP to an Android mobile device has been proposed as the initial case study:

- Customer management
- Items management
- Invoices management
- Orders management
- Shipment management

To this end, adapters have been developed for each ERP system-integrated.

The development of these adapters has been carried out on the NET WEB API platform [15] [16] [17] [27]. Communication with each ERP has been done with SOAP web services in Microsoft Dynamic AX [25] and NAV [24] and using DI API in SAP Business One. [26] .The information that has been transferred covers the following information entities: Customers, Items, Orders, Invoices and Shipment. Information Correlations of each ERP to the established standard have been made. Figure 3 shows one of the tables with the established correlations in the Customer entity.

NAVISION (NAV)	AXAPTA (AX)	SAP BUSINESS ONE	STANDARD
No	AccountNum	cardcode	numero
Name	Name	CardName	nombre
Name2	NameAlias	CardFname	nombre2
Address	Address	Address	direccion
Address	Address[2]	----	direccion2
City	City	MailCity	ciudad
Phone No	[Type ='Phone']/Locator	Phone1	telefono
Price including Vat	InclTax	-------	incluyeIva
VAT Registration No	VATNum	VATRegNum	numeroIva
Gen. Bus. Posting Group		----	genBusPostingGoup
PostCode	ZipCode	ZipCode	codigoPostal
Debit Amount		Balance	cantidadDebito
Credit Amount		----	cantidadCredito
E-Mail	[Type ='Email']/Locator	e_mail	eMail
No. Series	NumberSequenceGroup	----	numeroSecuencia
VAT Bus. Posting Group	Tax Group	VATGroup	grupoImpuestos
No. of Orders		---	numeroPedidos
No. of Invoices		-----	numeroFacturas
Customer Posting Group	CustGroup	GroupCode	grupoCliente

Fig. 3. Information Correlations

Each ERP adapters provides a common service interface in communication services with mobile devices. The application will communicate with the different ERP setting the adapter configuration with the one you want to connect, then it will perform the same operations regardless of the ERP at the other end of the communication.

For example, if we want to perform a search for an item, the following request will be made to the adapter, independently of the adapter involved.

POST request with the following JSON data to web service "/sap/item /search ' adapter in this case SAP Business One.

```
{"Cri-
tria":{"Criterion1":"Value1","Criterion2":"Value2",…,"Cri
terionN":"ValueN"}},
{"Filters":["Filter1","Filter2","Filter3",…,"FilterN"]}
```

The criteria indicates the record being searched and filters the information that you want to retrieve of that record. The response obtained after doing the request to this service shall comprise the information specified by the filters of all items that match the criteria sent Figure 4.

The developed application has a set of modules corresponding to the entities and operations of ERP mobilized Figure 5 A). Can be made in different ERP systems management operations such as search items Figure 5 B), search invoices Figure 5 D) orders creation Figure 5 C) and so on.

The mobile application offers a configuration module Figure 6. C) and D) where you must specify the ERP system which connected to the application at each moment. Options such as barcode scanning and creating PDF documents Figure 6 A) and B) have been developed as examples of the features that can be added by taking advantage of characteristics of mobile devices.

```
{"LISTA":{"ENTIDAD":[
{
    "numero": "0110101",
    "descripcion": "CEBADA",
    "precioUnidad": "128.020000",
    "inventario": "56.128343",
    "grupoIvaArticulo": "R3",
    "tamanoLote": "1.000000"
},
{
    "numero": "0110102",
    "descripcion": "TRIGO",
    "precioUnidad": "236.000000",
    "inventario": "85.033667",
    "grupoIvaArticulo": "R3",
    "tamanoLote": "1.000000"
},
```


Fig. 4. Search response

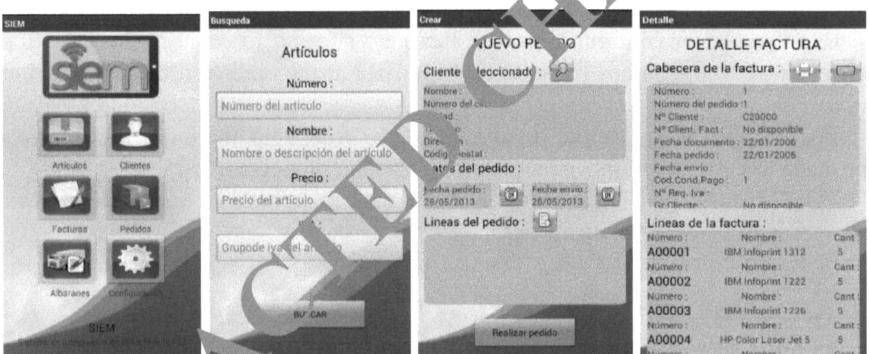

Fig. 5. A) Home menu B)Item Search C) New Order D)Bill Detail

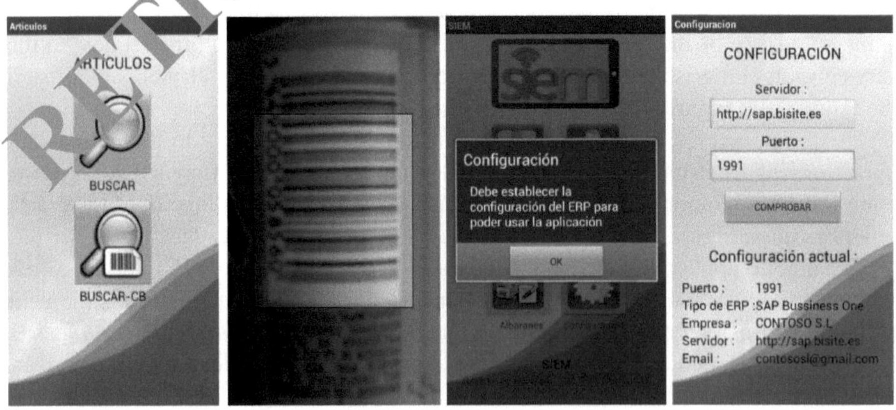

Fig. 6. A) Item Codebar Search B) Codebar Scan C) & D) Configuration Module

The developed application is able to perform the operations described below in integrated ERP systems. To perform the different operations which are available on the adapter, simply connect the adapter and set the adapter configuration imposed. Having established the configuration you can access the various modules of the application and perform the operations that enable the adapter.

5 Conclusions and Future Works

This developed system allows transferring characteristics and pre-auto-sales tasks to mobile Android devices with different ERP systems on the market such as Microsoft Dynamics AX, Microsoft Dynamics NAV and SAP Business One. Creating a common standard of information enables to establish in the same way communication between the mobile device and any adapter. This facilitates the development of applications for different ERP systems. The process of defining the standard of information has been a complicated process that has required an exhaustive research to locate the different ERP information and its correlation with the standard. The process of publishing the information in different ERP is varied. In Microsoft ERPs it is done through a series of SOAP web services to be configured and to provide remote access publishing. In the case of SAP access to information is done by DI API. This process is simplified with the established standard because access to information is common for ERPs. An application that uses this developed system has been carried out in Android's platform. This makes possible a subsequent development of applications in other platforms such as IOS, Blackberry OS or Windows Phone, even in web applications.

Increasing adapter's usefulness is proposed as a future project [14] [13] [12] [11]. Currently, the developed adapters only cover a small usefulness of every possible operations of the ERP. Thanks to the platform where they have been created, more services can be added later to different functional modules of the ERP. This will allow increasing the usefulness of the developed application or even develop other applications related with other functional module of the ERP.

References

1. Verdelho Azevedo, P., Romão, M., Rebelo, E.: Advantages, Limitations and Solutions in the Use of ERP Systems (Enterprise Resource Planning) – A Case Study in the Hospitality Industry. 5, 264–272 (2012)
2. Johansson, B., Ruivo, P.: Exploring Factors for Adopting ERP as SaaS (9), 94 – 99 (2013)
3. Clemens, B., Cata, T., Hackbarth, G.: Mobile Device Considerations for Supply Chain and ERP Related Systems,16 pages (2012)
4. Al Bar, A., Mohamed, E., Khursheed Akhtar, M., Abuhashish, F.: A preliminary review of implementing Enterprise Mobile Application in ERP environment 11(04), 77–82 (2011)
5. Zur Muehlen, M., Nickerson, J.V., Swenson, K.D.: Developing web services choreography standards—the case of REST vs. SOAP (40), 9–29 (2005)
6. Agüero, J., Rebollo, M., Carrascosa, C., Julián, V.: MDD-Approach for developing Pervasive Systems based on Service-Oriented Multi-Agent Systems

7. Závodská, A., Šramová, V., Aho, A.-M.: Knowledge in Value Creation Process for Increasing Competitive Advantage. Advances in Distributed Computing and Artificial Intelligence Journal 1(3), 35–47 (2012)
8. Corchado, J.M., Fyfe, C.: Unsupervised neural method for temperature forecasting. Artificial Intelligence in Engineering 13(4), 351–357 (1999)
9. Fdez-Riverola, F., Corchado, J.M.: CBR based system for forecasting red tides. Knowledge-Based Systems 16(5), 321–328 (2003)
10. Tapia, D.I., Abraham, A., Corchado, J.M., Alonso, R.S.: Agents and ambient intelligence: case studies. Journal of Ambient Intelligence and Humanized Computing 1(2), 85–93 (2010)
11. Corchado, J.M., Lees, B.: Adaptation of cases for case based forecasting with neural network support. Soft Computing in Case Based Reasoning, 293–319 (2001)
12. Corchado Rodríguez, J.M.: Redes Neuronales Artificiales: un enfoque práctico. Servicio de Publicacións da Universidade de Vigo, Vigo (2000)
13. Bajo, J., Corchado, J.M.: Evaluation and monitoring of the air-sea interaction using a CBR-agents approach. In: Muñoz-Ávila, H., Ricci, F. (eds.) ICCBR 2005. LNCS (LNAI), vol. 3620, pp. 50–62. Springer, Heidelberg (2005)
14. Fraile, J.A., Bajo, J., Corchado, J.M., Abraham, A.: Applying wearable solutions in dependent environments. IEEE Transactions on Information Technology in Biomedicine 14(6), 1459–1467 (2011)
15. Corchado, J.M., De Paz, J.F., Rodríguez, S., Bajo, J.: Model of experts for decision support in the diagnosis of leukemia patients. Artificial Intelligence in Medicine 46(3), 179–200 (2009)
16. De Paz, J.F., Rodríguez, S., Bajo, J., Corchado, J.M.: Case-based reasoning as a decision support system for cancer diagnosis: A case study. International Journal of Hybrid Intelligent Systems 6(2), 97–110 (2009)
17. Tapia, D.I., Rodríguez, S., Bajo, J., Corchado, J.M.: FUSION@, a SOA-based multi-agent architecture. In: International Symposium on Distributed Computing and Artificial Intelligence (DCAI 2008), pp. 9–107 (2008)
18. Corchado, J.M., Aiken, J.: Hybrid artificial intelligence methods in oceanographic forecast models. IEEE Transactions on Systems, Man, and Cybernetics, Part C: Applications and Reviews 32(4), 307–313 (2002)
19. Corchado, J.M., Aiken, J., Rees, N.: Artificial intelligence models for oceanographic forecasting. Plymouth Marine Laboratory (2001)
20. Rodríguez, S., Pérez-Lancho, B., De Paz, J.F., Bajo, J., Corchado, J.M.: Ovamah: Multiagent-based adaptive virtual organizations. In: 12th International Conference on Information Fusion, FUSION 2009, pp. 990–997 (2009)
21. Tapia, D.I., Alonso, R.S., De Paz, J.F., Corchado, J.M.: Introducing a distributed architecture for heterogeneous wireless sensor networks. In: Omatu, S., Rocha, M.P., Bravo, J., Fernández, F., Corchado, E., Bustillo, A., Corchado, J.M. (eds.) IWANN 2009, Part II. LNCS, vol. 5518, pp. 116–123. Springer, Heidelberg (2009)
22. Rodríguez, S., de Paz, Y., Bajo, J., Corchado, J.M.: Social-based planning model for multiagent systems. Expert Systems with Applications 38(10), 13005–13023 (2011)
23. Pinzón, C.I., Bajo, J., De Paz, J.F., Corchado, J.M.: S-MAS: An adaptive hierarchical distributed multi-agent architecture for blocking malicious SOAP messages within Web Services environments. Expert Systems with Applications 38(5), 5486–5499
24. Studebaker, D.: Programming Microsoft® Dynamics™ NAV 2009. Packt Publishing Ltd., Birmingham (2009) ISBN 978-1-847196-52-1

25. Ansari, A.: Inside Microsoft Dynamics AX 2012. Microsoft Press, United States of America (2012) ISBN: 978-0-7356-6710-5
26. Niefert, W.: SAP® Business ONE Implementation. Packt Publishing Ltd., Birmingham (2009) ISBN 978-1-847196-38-5
27. Zeitler, A., Kheyrollahi, A.: Pro ASP.NET Web API: HTTP Web Services in ASP.NET (2013) ISBN-13: 978-1430247258
28. Tapia, D.I., De Paz, J.F., Rodríguez, S., Bajo, J., Corchado, J.M.: Multi-agent system for security control on industrial environments. International Transactions on System Science and Applications Journal 4(3), 222–226 (2008)
29. Borrajo, M.L., Baruque, B., Corchado, E., Bajo, J., Corchado, J.M.: Hybrid neural intelligent system to predict business failure in small-to-medium-size enterprises. International Journal of Neural Systems 21(04), 277–296 (2011)
30. De Paz, J.F., Rodríguez, S., Bajo, J., Corchado, J.M.: Mathematical model for dynamic case-based planning. International Journal of Computer Mathematics 86(9-11), 1719–1730 (2009)
31. Bajo, J., De Paz, J.F., Rodríguez, S., González, A.: Multi-agent system to monitor oceanic environments. Integrated Computer-Aided Engineering 17(2), 131–144 (2010)
32. Corchado, J.M., Bajo, J., De Paz, J.F., Rodríguez, S.: An execution time neural-CBR guidance assistant. Neurocomputing 72(13), 2743–2753 (2009)
33. Panorama Consulting. Panorama Consulting, http://panorama-consulting.com/
34. Jitterbit, http://www.jitterbit.com/
35. JSON, http://www.json.org/
36. VTD-XML, http://vtd-xml.sourceforge.net/persistence.html
37. https://itunes.apple.com/es/app/sap-business-one/id392606876?mt=8
38. https://play.google.com/store/apps/details?id=b1.mobile.android&hl=es_419
39. http://www.windowsphone.com/es-es/store/app/sap-b1-mobile-apps/d6d482e3-b1e1-40f0-8f59-89c52b2df4c6

RETRACTED CHAPTER: Prediction System of Pollen Allergies in Mobile Devices

Daniel Hernández[1], Ana de Luis[1], and Sigeru Omatu[2]

[1] Departamento de Informática y Automática, Universidad de Salamanca
Plaza de la Merced, s/n, 37008, Salamanca, España
{danihiglesias,adeluis}@usal.es
[2] Department of Electronics, Information and Communication Engineering,
Osaka Institute of Technology, Osaka 535-8585, Japan
omatu@rsh.oit.ac.jp

Abstract. Although pollen allergies have a high incidence in society, it is not very common to use applications that provide data on pollen levels from different measuring points and also predict the allergies a user may experience. This paper introduces a system adapted to mobile devices that displays levels of pollen in the Spanish region of Castile and León in an easy way. The proposed system also processes the information provided by users about their health, and uses the historical data of pollen to detect and estimate allergies. The system incorporates an algorithm based on statistical tests to carry out the detection of allergies.

1 Introduction

Pollen allergy is a disease that affects 15% of the Spanish population, reaching as much as 30% of the young population [1] [2]. Currently, there are few applications for mobile terminals that can obtain data about the level of pollen in a given environment and predict allergies. However, it is possible to find similar and more well-known services, such as those providing information about the weather. Given the proliferate growth in the use of mobile devices[12][13][14][15][16], it would be of great interest to create an application that facilitates access to information about pollen levels and then use this information to predict allergies. An application of this kind can combine the information available in web repositories with the contextual information obtained from a mobile device [17][18][19][20][21][22][23][24][25]. Information about the pollen (historical and captured in-situ information) can be used in conjunction with the user's location and health status to detect and determine potential types of pollen allergy [3][26][27][28][29][30][31].

As mentioned above, there are some applications in Spain that provide information about pollen levels. These applications use the data provided by the Spanish Society of Allergology and Clinical Immunology [11]. Some of these applications are Polen Control, or AlertaPolen AlergoAlarm, each of which provides information on the pollen although they do not detect a user's allergies. This work shows an application developed in Android that provides access to information about pollen levels from mobile devices in a quick and easy way [32][33]. Furthermore, we present an automatic process for

---

The original version of this chapter was retracted: The retraction note to this chapter is available at https://doi.org/10.1007/978-3-319-07476-4_22

J.B. Perez et al. (eds.), *Trends in Practical Applications of Heterogeneous Multi-agent, Systems. The PAAMS Collection*, Advances in Intelligent Systems and Computing 293,
DOI: 10.1007/978-3-319-07476-4_5, © Springer International Publishing
Switzerland 2014, corrected publication 2024

detecting allergies, which makes use of the information indicated by the user about his or her health status and combines it with the location and historical data of pollen in order to predict allergies based on statistical tests. The execution of the algorithm is performed on a remote server to reduce the computational load on the mobile terminal. The system used the information of historical levels of pollen available in the open data portal of the Government of Castile and León [10].

The paper is organized as follows: section 2 includes a revision of related work, Section 3 describes our proposal, and finally sections 4 and 5 provide the preliminary results and conclusions obtained after testing the proposed approach.

2 Background

The field of medicine provides different tools to predict disease in certain areas, such as detection of asthma presented in [4], in which the authors used data from the health history of children aged 1-3 years to predict asthma. In the same line, there is an API [5] used for predicting infant and adult asthma [6] using data collected from patients between 2-21 years of age. Numerous studies [7] on the detection of pollen allergies indicate the incidence of the different grasses in the state of health of patients with an even greater level of detail than just a few years ago.

Today there are applications for mobile devices that visualize the levels of pollen by using measurement information centers. For example Figure 1 shows the Pollen Alert app, which reports pollen levels in the Spanish towns where the measuring centers are located. The data are measured periodically and are provided by a network of centers including the Aerobiology Committee of the Spanish Society of Allergology. The Clinical Immunology (SEAIC) also supports this application.

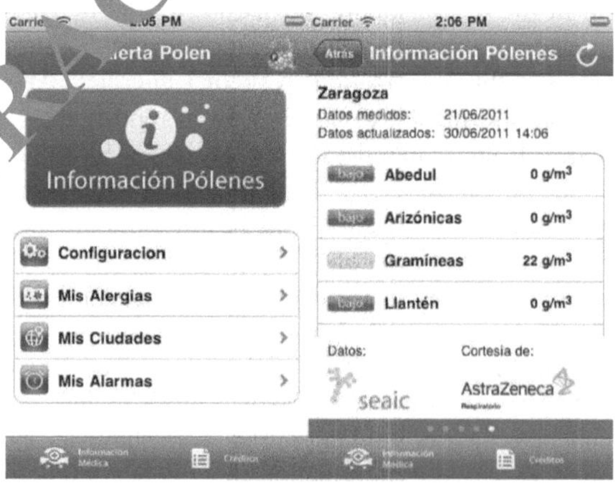

Fig. 1. Pollen Alert app

The Pollen Alert app provides measurement data from different station across Spain. Users may filter this information according to several criteria such as type of pollen (grasses, olive, Salsola, etc.) city or cities etc. Additionally, the application allows setting alarms to alert the users when the amount of pollen in the air from a certain type of grass exceeds a pre-established level of contamination.

Other applications such as Alergo Alarm were developed by SEAIC (Spanish Society of Allergology and Clinical Immunology) and Almirall (Spanish pharmaceutical company). These applications are more for primary care physicians. They provide daily information of levels in Spain and predict possible levels to assess the risks of allergies.

As with the previous application, Allergo triggers alarms based on predefined levels. The application filter information by type of grass and location. The filtering process reduces the information for display on the mobile.

In addition to Alarm Alergo, the Spanish Society of Allergology and Clinical Immunology and Almirall have other applications such as Pollen Control, which is used to monitor and diagnose allergies related to environmental pollen, especially in patients with hay fever. Pollen Control tracks the user symptomatic progression, thus allowing the user's allergist to obtain information about the relationship between the pollen levels and their evolution. In addition, this application also provides information on daily pollen levels, and predicts their expected evolution.

3 Proposed Reasoning System

The developed application uses the information provided by the Junta de Castille and [9] León about present and past levels of pollen. The current pollen levels are shown on a map (the google maps application is used for this purpose). This representation facilitates the selection of the nearest station to the user in an easy and intuitive way. Figure 2a shows a screenshot with information about pollen levels. The red circles represent the different types of pollen that are present with a medium or high level in the areas specifically requested by the user. In this way, the user can quickly and easily view the number of relevant levels in the measurement centers close to the area of observation. The proposed system obtains information from the measurement centers by selecting one of the markers in the map. The detailed information about this center is shown in a dialog box, which presents personalized information about the pollen in a more specific area, representing medium or high levels. In Figure 2b, the yellow and red circles represent medium and high levels of pollen respectively. The arrow on the right represents a prediction about the level of pollen (increases, reduces or remains stable).

Scrolling down the screen to the left displays the second screen. Figure 3 displays the different pollens that are available for filtering. By checking or unchecking the boxes pollens will appear or disappear on the google maps shown previously.

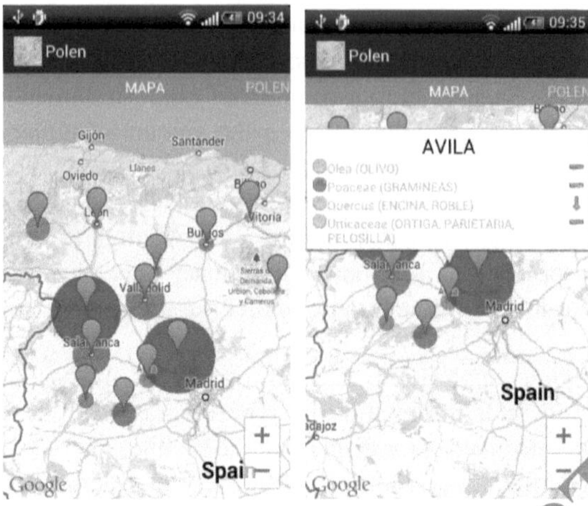

Fig. 2. a) Google maps with pollen levels in different seasons. b) Levels of pollen for the selected station.

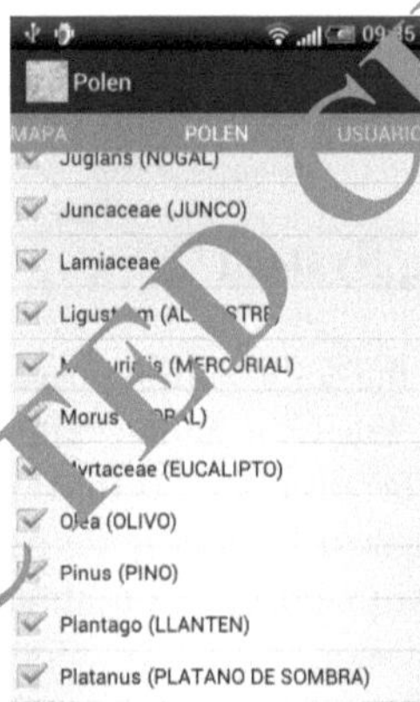

Fig. 3. Selectable pollens.

The third screen contains three tabs, although the last one only displays help information that it is not relevant in the development of our study. The first tab is used to add information about the health status of the user. The user has to indicate the date, the health status and the location associated to a measurement (once a week because it is the data update period). Figure 4a displays a screenshot of this information. In Figure 4b, the second tab contains the prediction for allergies, which is calculated using a Fisher's test according to the health status of the user and the historical values of pollen. [8]. Pollens are arranged according to allergy for each type of grass. The health status of the user and the level of the grass are displayed, making it easy to verify and validate the results. The information is processed on a remote server through a REST request. This way the mobile terminal does not execute the algorithm and the level of traffic is reduced because the weight of the historical data file is 1.5 MB. The current level file only takes around 25KB so the traffic is very small.

Fig. 4. a) User state. b) Predictions

4 Results

In this section we present a demonstration about the prediction system. The user indicates the date the measurements are taken, his or her health status and location; according to this information the system then provides a prediction of the types of grasses the user may be allergic to.

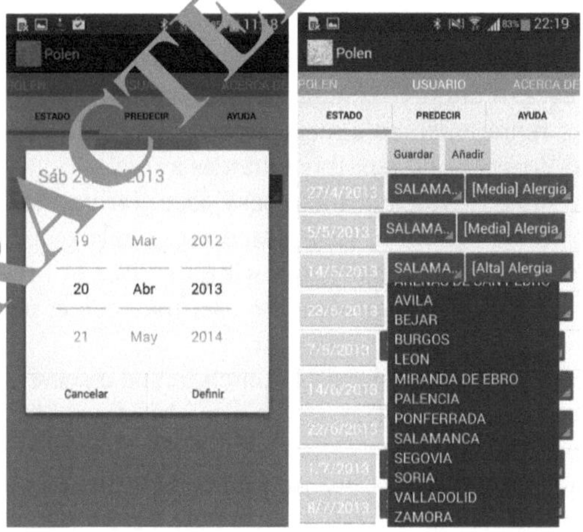

Fig. 5. a) Selecting a measurement date. b) Site selection of measurement.

As seen in Figure 5.a the user introduces the data about his or her health status, location and date. The values for location and date are not established automatically because the user can introduce previous values. The place is introduced according to the Figure 5.b.

Figure 6.a indicates the health status of the user for that date. There are three levels of intensity with these possible values: no allergy, medium allergy or high allergy. The first option removes the measurements. After entering this data, the user selects the "Predict" button and sends the data to the server, which in turn calculates the allergy and returns the results to the user.

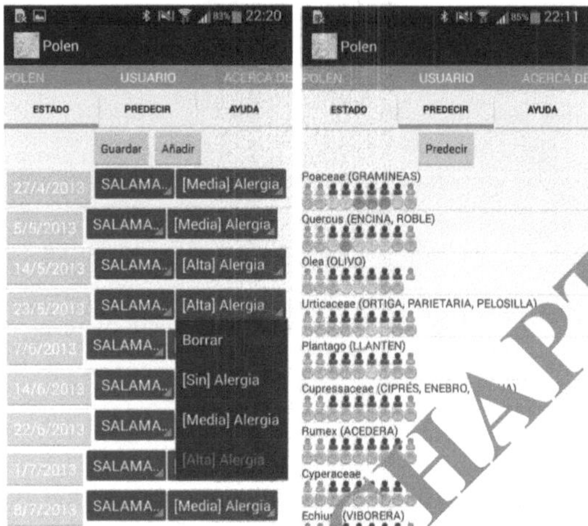

Fig. 6. a) Introduction level of allergy. b) Calculated results.

The user can then view the results in the mobile. The types of grass are sorted according to their relevance, as represented in Figure 6.b. In this figure we can see the different types of grass. For each one we have a history of the user's health status (green for a zero level of allergy, orange for average, and red for a high level) and a history of the grass represented as circles. According to the information, the user is probably allergic to Poaceae and possibly Quercus as well.

The presented system provides fast and easy access from mobile devices to the information of the level of pollen. A view based on google maps makes it possible to monitor the level of all zones globally, allowing the user to obtain detailed information about a specific region. Furthermore, the user may detect possible allergies using the historical data.

The application has two different parts: a mobile and a server. The mobile application was developed for Android. The server was also developed using java web services REST (Representational State Transfer) and JSON (JavaScript Object Notation) for sending the information. The software is available in the play store of Android, and the app is free and has no advertising. It may be downloaded by searching the name "APCYL: Alergia al polen CyL".

References

1. Caillaud, D., Toloba, Y., Raobison, R., Besancenot, M.: Health impact of exposure to pollens: A review of epidemiological studies, Revue des Maladies Respiratoires (2013)
2. Subiza, J.: Allergenic pollens in Spain. Allergologia et Immunopathologia 32(3), 121–124 (2004)

3. De Benito, V., Menchaca, J.M., Rubio, M.C., Sánchez, Y., Rodríguez, B., Soto, J.: Identificación de los taxones de pólenes alergénicos en pacientes polínicos para conocer la temporada de riesgo. Allergologia et Immunopathologia 32(34), 228–232 (2004)
4. Pescatore, A., Dogaru, C., Duembgen, L., Silverman, M., Gaillard, E., Spycher, B., Kuehni, C.: A simple asthma prediction tool for preschool children with wheeze of cough. Journal of Allergy and Clinical Immunology 133(1), 111–118 (2014)
5. Leonardi, N., Spycher, B., Strippoli, M.: Validation of the Asthma Predictive Index and comparison with simpler clinical prediction rules. Journal of Allergy and Clinical Immunology 127(6), 1466–1472 (2011)
6. Balemans, W., van der Ent, C., Schilder, A.: Prediction of asthma in young adults using childhood characteristics: Development of a prediction rule. Journal of Clinical Epidemiology 59(11), 1207–1212 (2006)
7. Lewis, M., Leonard, B.: Digital Pathology Techniques in Pollen Assessment, Journal of Allergy and Clinical Immunology. Journal of Allergy and Clinical Immunology 131(2), AB77 (2013)
8. Requena, F., Martín Ciudad, N.: A major improvement to the Network Algorithm for Fisher's Exact Test in 2×c contingency tables. Computational Statistics & Data Analysis 51(2), 490–498 (2006)
9. http://www.saludcastillayleon.es/ciudadanos/es/polen
10. http://www.jcyl.es
11. http://www.seaic.org
12. Corchado, J.M., Fyfe, C.: Unsupervised neural method for temperature forecasting. Artificial Intelligence in Engineering 13(4), 351–357 (1999)
13. Fdez-Riverola, F., Corchado, J.M.: CBR based system for forecasting red tides. Knowledge-Based Systems 16(5), 321–328 (2003)
14. Tapia, D.I., Abraham, A., Corchado, J.M., Alonso, R.S.: Agents and ambient intelligence: case studies. Journal of Ambient Intelligence and Humanized Computing 1(2), 85–93 (2010)
15. Corchado, J.M., Lees, B.: Adaptation of cases for case based forecasting with neural network support. Soft computing in case based reasoning, 293–319 (2001)
16. Corchado Rodríguez, J.M.: Redes Neuronales Artificiales: un enfoque práctico. Servicio de Publicacións da Universidade de Vigo, Vigo (2000)
17. Bajo, J., Corchado, J.M.: Evaluation and monitoring of the air-sea interaction using a CBR-agents approach. In: Muñoz-Ávila, H., Ricci, F. (eds.) ICCBR 2005. LNCS (LNAI), vol. 3620, pp. 50–62. Springer, Heidelberg (2005)
18. Fraile, J.A., Bajo, J., Corchado, J.M., Abraham, A.: Applying wearable solutions in dependent environments. IEEE Transactions on Information Technology in Biomedicine 14(6), 1459–1467 (2011)
19. Corchado, J.M., De Paz, J.F., Rodríguez, S., Bajo, J.: Model of experts for decision support in the diagnosis of leukemia patients. Artificial Intelligence in Medicine 46(3), 179–200 (2009)
20. De Paz, J.F., Rodríguez, S., Bajo, J., Corchado, J.M.: Case-based reasoning as a decision support system for cancer diagnosis: A case study. International Journal of Hybrid Intelligent Systems 6(2), 97–110 (2009)
21. Tapia, D.I., Rodríguez, S., Bajo, J., Corchado, J.M.: FUSION@, a SOA-based multi-agent architecture. In: International Symposium on Distributed Computing and Artificial Intelligence (DCAI 2008), pp. 99–107 (2008)

22. Corchado, J.M., Aiken, J.: Hybrid artificial intelligence methods in oceanographic forecast models. IEEE Transactions on Systems, Man, and Cybernetics, Part C: Applications and Reviews 32(4), 307–313 (2002)
23. Corchado, J.M., Aiken, J., Rees, N.: Artificial intelligence models for oceanographic forecasting. Plymouth Marine Laboratory (2001)
24. Rodríguez, S., Pérez-Lancho, B., De Paz, J.F., Bajo, J., Corchado, J.M.: Ovamah: Multiagent-based adaptive virtual organizations. In: 12th International Conference on Information Fusion, FUSION 2009, pp. 990–997 (2009)
25. Tapia, D.I., De Paz, J.F., Rodríguez, S., Bajo, J., Corchado, J.M.: Multi-agent system for security control on industrial environments. International Transactions on System Science and Applications Journal 4(3), 222–226 (2008)
26. Borrajo, M.L., Baruque, B., Corchado, E., Bajo, J., Corchado, J.M.: Hybrid neural intelligent system to predict business failure in small-to-medium-size enterprises. International Journal of Neural Systems 21(04), 277–296 (2011)
27. De Paz, J.F., Rodríguez, S., Bajo, J., Corchado, J.M.: Mathematical model for dynamic case-based planning. International Journal of Computer Mathematics 86(10-11), 1719–1730 (2009)
28. Tapia, D.I., Alonso, R.S., De Paz, J.F., Corchado, J.M.: Introducing a distributed architecture for heterogeneous wireless sensor networks. In: Omatu, S., Rocha, M.P., Bravo, J., Fernández, F., Corchado, E., Bustillo, A., Corchado, J.M. (eds.) IWANN 2009, Part II. LNCS, vol. 5518, pp. 116–123. Springer, Heidelberg (2009)
29. Rodríguez, S., de Paz, Y., Bajo, J., Corchado, J.M.: Social-based planning model for multiagent systems. Expert Systems with Applications 38(10), 13005–13023 (2011)
30. Pinzón, C.I., Bajo, J., De Paz, J.F., Corchado, J.M.: S-MAS: An adaptive hierarchical distributed multi-agent architecture for blocking malicious SOAP messages within Web Services environments. Expert Systems with Applications 38(5), 5486-5499
31. Corchado, J.M., Bajo, J., De Paz, J.F., Rodríguez, S.: An execution time neural-CBR guidance assistant. Neurocomputing 72(13), 2743–2753 (2009)
32. Griol, D., García-Herrero, J., Molina, J.M.: Combining heterogeneous inputs for the development of adaptive and multimodal interaction systems. Advances in Distributed Computing And Artificial Intelligence Journal
33. Serrano, E., Gómez-Sanz, J.J., Botía, J.A., Pavón, J.: Intelligent data analysis applied to debug complex software systems. Neurocomputing 72(13), 2785–2795

A Web-Based Agent-Oriented Approach to Address Heterogeneity in Cooperative Embedded Systems

Jean-Paul Jamont[1], Lionel Médini[2], and Michael Mrissa[2]

[1] Université de Grenoble Alpes, Laboratoire LCIS, Valence, France
jean-paul.jamont@lcis.grenoble-inp.fr
[2] Université de Lyon, LIRIS - UFR FST, CNRS, 43, boulevard du 11 novembre 1918, 69622 Villeurbanne cedex
{michael.mrissa,lionel.medini}@univ-lyon1.fr

Abstract. This paper presents a MAS-oriented approach to enable emergence and execution of complex functionality among a fleet of heterogeneous connected objects. It relies on the Web of Things paradigm, in which such objects communicate using Web standards. In order to homogenize the objects and extend their capabilities, our approach is based on agents that can be deployed either on objects or in the cloud. Such agents can embody the object behaviors and perform negotiation to achieve collaborative functionalities.

Keywords: avatars, multi-robot systems, embedded systems, Web of things.

1 Introduction

Developments in wireless technologies have strongly impacted embedded systems. Embedded systems have become sets of small interconnected heterogeneous devices strongly related to their environment. Devices need to cooperate when they do not have certain necessary skills or knowledge to accomplish their individual goals but also to meet objectives of the overall system. Such systems can be then observed at two levels: an individual level (the device layer) and a social level (the global system).

Designing a cooperative embedded systems require to meet numerous challenges like heterogeneity, scalability, openness etc. Heterogeneity means the diversity of the devices in terms of platform, connectivity, operating systems etc. A device can be a complex robots reasoning with abstract symbolic representations of its own environment, a mobile phone, a hard resources constrained sensors or chipless things like a RFID tagged books. The heterogeneity management, which is the ability to describe data and services exposed by the devices, is a hard point.

In such applications, multiagent systems (MAS) are used because they offer many relevant models to implement collective and adaptive behaviors. The

J.B. Perez et al. (eds.), *Trends in Practical Applications of Heterogeneous Multi-agent Systems. The PAAMS Collection*, Advances in Intelligent Systems and Computing 293,
DOI: 10.1007/978-3-319-07476-4_6,

contribution of this paper is a Web-based agent-oriented approach to address the previously exposed challenges. This approach is based on the concept of avatar which is a Web abstraction of a real world device. According to a given Avatar architecture, the result is an open, generic and Web-based solution to deliver high-level, user-understandable functionality, while interacting with a set of various physical objects, enabling cooperation between them.

Section 2 the background of this work i.e. works integration of heterogeneous agents in a same MAS using Web. Section 3 describes our decentralized approach and defines the notion of avatar, before illustrating the interest of our approach in the context of the WoT with a motivating scenario. Before to conclude, we introduce in Section 4 the internal architecture of avatar agents.

2 Background

The integration of heterogeneous agents in a same MAS is the subject of numerous studies since the late 90s. FIPA[1] has proposed many specifications to address the problem as the standardization of interactions. FIPA proposes some standard to treat heterogeneity as a communication problem(FIPA-ACL), from an architectures point of view (Agent Abstract Architecture), as a middleware integration problem etc.

The work of the most popular standardization in IT and most universally prevalent are those of the Web. It is therefore natural to focus on these standards to ensure interoperability of heterogeneous agents.

Works at the intersection of MAS and WS can be divided into three categories [1]. The first one concerns the use of a MAS as a mediator in the WS functional model like in [2,3,4,5]. The last categories use WS to make available the MAS through the Web according two different approaches :

- An integrated approach: WS are developed following an agent model to perform complex tasks such as management commercial transactions or interactions [6,7] or, on the other hand, WS are accessible through a multiagent framework [8,9].
- A decoupled approach: Starting from a given MAS, a WS layer enable agents to interact together according to Web interfaces [1,10,11,12].

3 Our Approach

We provide a decentralized approach that enables proactive and cooperative intelligence through the concept of avatar. We then explain how avatars enable collective intelligence. We then illustrate our approach through.

Extending Objects with Avatars. To introduce the concept of avatar, we need to define *object* and *proxy*.

[1] Foundation for Intelligent Physical Agents - `http://www.fipa.org/`

Objects Applications developed for the WoT put into play many heterogeneous. These objects are hardware or software entities. Establishing a relationship between these different types of entities creates the added-value the WoT brings to the user community. We can consider an object as a 4-tuple $O =< G, B, K, S >$, where $G = \{G_1, G_2, ..., G_n\}$ is its set of goals (what the object want fulfill), $K = \{K_1, K_2, ..., K_n\}$ its knowledge i.e. the set of re-usable abstractions about its environment and about other devices, $S = \{S_1, S_2, ..., S_n\}$ is the set of device's features and $B = \{B_1, B_2, ..., B_n\}$ its behavior i.e. a set of rules to define the logic of actions/reactions in response to internal/external stimuli.

We identify three types of physical objects:

1. Complex Objects: These objects provide software services and embed a Web server that offers service interfaces. It is then often trivial to link these objects together or with other software services. The only difference with traditional WS is the connection to physical objects that interact with their environment.
2. Lightweight Objects: These objects cannot embed Web servers due to restricted computing capacity (low memory, limited energy resources, restricted bandwidth) but it is often easy to link them to proxies. A proxy can embed a Web server. The (object,proxy) couple can be seen as a complex object that is physically distributed.
3. Bare Objects : These objects are passive objects that can be detected such as pallets with RFID (Radio Frequency IDentifier) tags. When such an object is in the range of a RFID reader, the reader receives a byte array. A logical link can then be established between the physical object and the byte array.

Proxy. In our context, a proxy is a projection of a physical object into the Web. Concretely, it is a Web intermediary for requests from clients seeking resources from other objects. We can define a proxy as a 2-tuple $P =< Kp, Sp >$ with $Kp \subseteq K$ and $Sp \subseteq S$. The selection of what is exposed depends of different strategies (energy management , privacy preserving, etc.).

Avatar. Our approach consists in extending objects with a virtual representation on the Web (Fig. 1). We call such an representation an avatar. Avatars are not simple proxies. An avatar is an autonomous entity (i.e. an agent) which has it own 4-tuple $A =< Ga, Ba, Ka, Sa >$ with $G \subset Ga$, $B \subset Ba$, $K \subset Ka$ and $S \subset S_a$. The increase of its knowledge and its skills comes from (1) the Web which is the avatar environment (so an avatar can access to the Web of data and WS) and (2) others avatars. Through their avatars, physical objects can be in interaction and particularly in cooperation.

Considering a bare object "pot of yogurt", its avatar should be able to identify its location and state by accessing and analyzing the history of crossed RFID readers. The avatar can therefore try to interact with another avatar that extends a temperature sensor at the same place, to determine whether or not the yogurt pot could be degraded. This example shows that more than having extended the capacities of things, we have, individually and globally, enriched their behavior.

Interactions between avatars can lead to exhibit collective behaviors. We can concretely reuse all the works exhibited in MAS in the context of avatars : an avatar also is an autonomous agent.

In order to plan how to achieve collaborative functions, avatars must negotiate with one another. This requires both a negotiation model and a communication protocol. These well-known models of the multiagent community are not developed here ([13,14]).

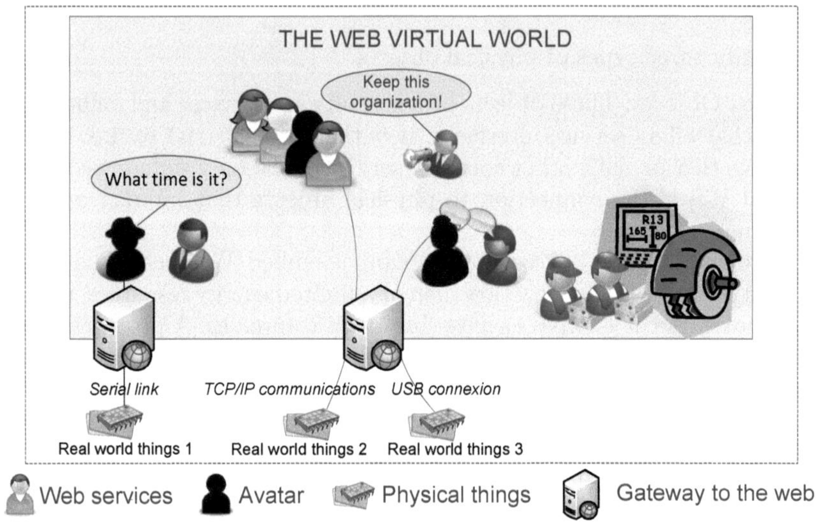

Fig. 1. Using avatars in the Web

Application of Our Approach. To illustrate our approach, we consider the following production chain scenario (inspired by [15]) (fig. 2).

We have to manage the flow of yogurt produced in a firm. These goods are packed in boxes. Boxes are stacked on RFID tagged plastic pallet. When a pallet is filled to its maximum capacity, a human operator places the pallet in the production deposit area. The pallets will now be handled by mobile robots. Pallets of yogurt have to be stored in any storage inside the warehouse. We assume that a yogurt should never stay more than 10 minutes in a place where the temperature is higher than 10°C, otherwise it becomes inappropriate for human consumption and an alternative scenario must be considered (i.e. recycling process or destruction).

Our approach leads to build WoT adaptive applications. We illustrate our approach in the context of the previously i scenario.

1. Production chain produces a pallet of yogurt: a pallet leaves the production line with a load of yogurt. It is deposited in the *production deposit area.*

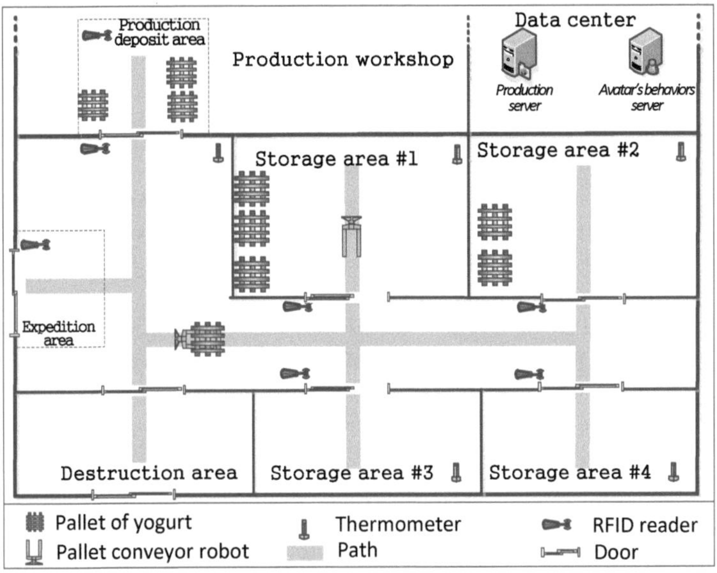

Fig. 2. Illustration of the scenario

   (a) The pallet RFID tag is read by the deposit area RFID reader.
   (b) The *avatar creation service* queries the *production server* to know what is the associated thing.
   (c) The *production server* informs it is a pallet of yogurt.
   (d) The *avatar builder service* downloads from the *code repository* the behavior associated to an abstraction of pallet of yogurt and generates an instance of avatar for the yogurt pallet.
   (e) The *avatar builder service* creates an *avatar* `(rfid_tag="#50 41 4c 4c 45 54 31 32 33 34 35 36 37 38", type="#Pallet", parameter_list=("#Yogurt") )`.
   (f) The association between the avatar's url (`#yogurt_pallet_1`) is memorized in the *context server*.
2. The pallet is deposited in *Storage area 1*.
   (a) The tag is read by the *RFID reader* device associated to *storage area1*.
   (b) Treatment of the tag by the RFID reader (`#RFID_reader_1`) avatar associated to this room.
   (c) The contextual event `You enter in #StorageArea_1` is send by the avatar `#RFID_reader_1` to the avatar `#yogurt_pallet_1`.
3. Avatar `#yogurt_pallet_1` has an introspection : it inspect its behaviour to find rules linked to a room modification.
   (a) He searches rules linked to the event `room modification`
      i. He searches rules linked to the instance `#StorageArea_1`
      ii. No rules are found then he search rules linked to concept behaviour introspection `#Area`
      iii. No rules are found then he search rules linked to super-concepts of `#Area` (here it is #Place).

iv. A rule is founded : `if temperature(#Place) > 10°  then "find a solution to be sure the storage period at this temperature is under 10 minutes"`.

(b) Application of rules:

i. The `#yogurt_pallet_1` avatar have a behaviour introspection to try to find functionality `getTemperature()`

ii. The function is not found then the avatar search another one which can provide functionality `getTemperature(p : place = #StorageArea_1)`.

iii. The thermometer avatar associated to `#StorageArea_1` inform it is able to provide this functionality.

iv. The pallet of yogurt avatar request functionality `getTemperature (#StorageArea_1)` of `#temperature_sensor_StorageArea_1`

v. The result is 12^oC then the pallet of yogurt avatar applies rule `search a solution`. The 1st solution is `self-move : move(p : pallet = #Pallet_1)`

vi. Introspection to find functionality `move(p : pallet = #Pallet_1)`.

vii. The functionality is not found then the avatar search another one which can provide functionality `move(thing : pallet, from : place, to : place )`

viii. Avatars `#Pallet_conveyor_1` and `#Pallet_conveyor_2` inform they can provide this functionality. `#Pallet_conveyor_3` has not answer because if is under maintenance.

ix. Avatars `#Yogurt_pallet_1`, `#Pallet_conveyor_1` and `#Pallet_conveyor_2` negotiate together : `#Pallet_conveyor_1` proposes the best quality of service because it can supply the service earlier. In fact, it is free of load contrary to `#Pallet_conveyor_2`.

x. `#Yogurt_pallet_1` requires this functionality of `#Pallet_conveyor_1`

xi. `#Pallet_conveyor_1` accepts.

4. Transportation of `#Yogurt_pallet_1`

(a) ...

This scenario shows how objects can benefit from a Web-based and agent-oriented approach that extends the interactions possibilities and enhance their life cycle, allowing better and natural integration into their environment.

4 Avatar Architecture

The avatar architecture is referred to as "WoT Runtime Environment", as depicted in figure 3. It is composed of a framework in which are plugged a set of components called managers, which are grouped in modules. Each manager takes charge of a specific concern and interacts with other components using a specific API. The core module of the avatar architecture allows deploying the components into this framework and furnishes low-level components that are required by several managers, to perform caching and reasoning tasks. The applicative code that implements the functionalities that the objects can realize

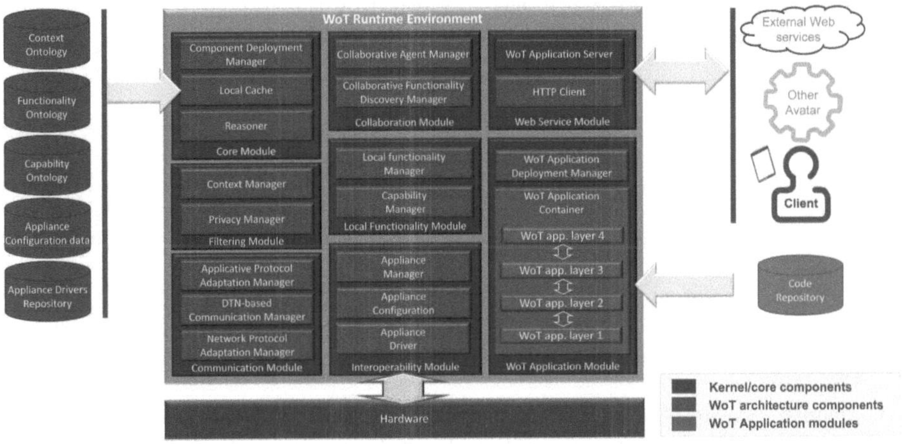

Fig. 3. Architecture of an avatar

is dynamically deployed in a second internal framework, called the "WoT Application Container". Each logical component of the avatar architecture can be executed either on the object or in the cloud.

The "Collaborative Agent Manager" is responsible of the MAS aspects of the avatar architecture and embeds the communication and negotiation models. It queries the "Collaborative Functionality Discovery Manager" to retrieve a list of potential functionalities that cannot be locally achieved by an object, but require collaboration. It interacts with other avatars through the Web service module: the "WoT Application Server" component exposes both the available functionalities on each object and the negotiation protocol as REST resources; each avatar thus uses the HTTP client module to query these resources in the WoT infrastructure.

5 Conclusion

We proposed an approach to enable cooperative intelligence for the Web of Things. Our approach is decentralized, Web-based and agent-oriented to enable proactive and cooperative intelligence between heterogeneous objects with the help of Web-based languages and protocols. Its builds on an agent abstraction called avatar that extends a real world device on the Web.

Implementing avatars according the introduced architecture enables to demonstrate the applicability of our work in the context of a production chain scenario. The result is an open, generic and Web-based solution to deliver high-level, user-understandable functionality, while interacting with a set of various physical objects through their avatars, enabling cooperation between them.

As future work, we envision to develop MAS protocols between avatars, and to extend our architecture to make it adaptable to different aspects such as MANET environments and scalability problems.

Acknowledgment. This work was supported by the french *Agence Nationale de la Recherche* through contract <ANR-13-INFR-012>.

References

1. Melliti, T., Haddad, S., Suna, A., Fallah-Seghrouchni, A.E.: Web-masi: Multi-agent systems interoperability using a ws based approach. In: Proc. of the IEEE/WIC/ACM Int. Conf. on Intelligent Agent Technology, pp. 739–742. IEEE (2005)
2. Richards, D., Splunter, S., Brazier, F.T., Sabou, M.: Composing web services using an agent factory. In: Extending Web Services Technologies. Multiagent Systems, Artificial Societies, and Simulated Organizations, vol. 13, pp. 229–251. Springer, US
3. Maamar, Z., Mostéfaoui, S.K., Yahyaoui, H.: Toward an agent-based and context-oriented approach for ws composition. IEEE Trans. Knowl. Data Eng. 17(5), 686–697 (2005)
4. Morais, A.J.: A multi-agent approach for web adaptation. In: Demazeau, Y., Pavón, J., Corchado, J.M., Bajo, J. (eds.) 7th International Conference on Practical Applications of Agents and Multi-Agent Systems (PAAMS 2009). AISC, vol. 55, pp. 349–355. Springer, Heidelberg (2009)
5. Gutiérrez-García, J.O., Corchado, F.F.R., Koning, J.-L.: An obligation-based framework for web service composition via agent conversations. WIAS 10(2), 135–150 (2012)
6. Jin, T., Goschnick, S.: Utilizing web services in an agent based transaction model (abt). In: Workshop on Web Services and Agent-Based Engineering (2003)
7. El Falou, M., Bouzid, M., Mouaddib, A.-I., Vidal, T.: A distributed multi-agent planning approach for automated web services composition. WIAS 10(4), 423–445 (2012)
8. Ma, B., Chen, B., Bai, X., Huang, J.: Design of bdi agent for adaptive performance testing of web services. In: Proceedings of the 10th International Conference on Quality Software, pp. 435–440. IEEE Computer Society (2010)
9. Coria, J.A.G., Castellanos-Garzón, J.A., Corchado, J.M.: Intelligent business processes composition based on multi-agent systems. Expert Syst. Appl. 41(4), 1189–1205 (2014)
10. Ramírez, E.H., Brena, R.F.: Multi-agent systems integration in enterprise environments using web services. IJIIT 2(3), 72–88 (2006)
11. Goschnick, S., Balbo, S., Sterling, L., Sun, C.: Tandem - a design method for integrating web services into multi-agent systems. In: 5th Int. Joint Conf. on Autonomous Agents and Multiagent Systems (AAMAS 2006), pp. 939–941. ACM (2006)
12. Karaenke, P., Schuele, M., Micsik, A., Kipp, A.: Inter-organizational interoperability through integration of multiagent, web service, and semantic web technologies. In: Fischer, K., Müller, J.P., Levy, R. (eds.) ATOP 2009 and ATOP 2010. LNBIP, vol. 98, pp. 55–75. Springer, Heidelberg (2012)
13. FIPA: Review of fipa specifications (December 2006)
14. Weiss, G. (ed.): Multiagent Systems, 2nd edn. MIT Press, Cambridge (2013)
15. Gautier, P., Gonzalez, L.: L'Internet des objets: Internet, mais en mieux. AFNOR (2011)

RETRACTED CHAPTER: Multi-agent System for Occupational Therapy

Amparo Casado[1], Amparo Jiménez[1], Javier Bajo[2], and Sigeru Omatu[3]

[1] Universidad Pontificia de Salamanca,
C/Compañía 5, 37002, Salamanca, Spain
{acasadome,ajimenezvi}@upsa.es
[2] DIA, Universidad Politécnica de Madrid
ETS Ingenierios Informáticos, Campus Montegancedo, Boadilla del Monte, 28660, Spain
javier.bajo@upm.es
[3] Faculty of Engineeging, Osaka Institute of Technology
5-16-1 Omiya ahahi-ku Osaka, 535-8585, Japan
omatu@rsh.oit.ac.jp

Abstract. This paper presents a new solution for improving evaluation and intervention processes carried out by therapist. The proposed solution is based on a multiagent system especially designed for occupational therapy. The proposed solution aims at improving the traditional system to fulfill valuation scales, generation of therapy reports and extraction of statistics. Besides, the system proposes new techniques to improve the interaction with the user, by means of interactive cards that can be used for rehabilitation when working with intellectual handicapped patients. The paper shows the proposed application, as well as some preliminary results.

Keywords: Occupational therapy, evaluation, intervention, multi-agent systems.

1 Introduction

Occupational therapy evaluates the ability of a person to perform activities of daily living and intervenes if such capacity is at risk or threatened for any reason. Occupational therapists are the professionals responsible for carrying out, by hand, the essential processes of assessment and involvement of users in occupational centers [16, 10]. The current methodology has some problems in both processes: the difficulty of managing the vast amount of information generated on paper, the availability of this information, the lack of safety data, and some little update audiovisual media, etc. [16, 10]. Multi-agent systems and intelligent mobile devices architectures are suitable to handle complex and highly dynamic problems [2]. Agents and multi-agent systems have been successfully implemented in areas such as e-commerce, medicine, oceanography, robotics, etc. [3. 5, 7, 8, 9, 11, 13, 14, 15, 16, 18, 20, 21]. They have been recently explored as supervision systems, with the flexibility to be implemented in a wide diversity of scenarios, including health care environments. The current application of multi-agent systems in social environments is an area of increasing interest [11, 22, 23, 25, 28]. In general, the multi-agent system represents an appropriate

The original version of this chapter was retracted: The retraction note to this chapter is available at https://doi.org/10.1007/978-3-319-07476-4_22

J.B. Perez et al. (eds.), *Trends in Practical Applications of Heterogeneous Multi-agent, Systems. The PAAMS Collection*, Advances in Intelligent Systems and Computing 293,
DOI: 10.1007/978-3-319-07476-4_7,

approach for solving inherently distributed problems, whereby clearly different and independent processes can be distinguished.

This paper aims to improve and computerize the processes of assessment and intervention carried out by therapists, obtaining a mobile application that can help the medical staff to carry out their duties. This improves the traditional process, allowing the completion of rating scales, the generation of reports of therapy and extraction of statistics. It also improves the interaction between professionals and users using interactive maps that allow rehabilitation of mental disabilities and / or intellectual. Thus, we improve the evaluation process using scales implemented in the application and allowing better management and assessment of records and facilitating the completion of the scales and the measurement of users. We assist the intervention process by using flashcards, which improves user interaction and therapist, providing access to different types of issues and improving the user experience. We also facilitate the mobility of the therapist using a single mobile device to make the whole process of occupational therapy anywhere and by providing access to all information without depending on the location of the user or therapist.

The reminder of the paper is organized as follows: Section 2 reviews the state of the art, Section 3 introduces the proposed system and Section 4 presents the results and conclusions obtained.

2 Background

The Spanish Association of Professional Occupational Therapists (APETO) defines occupational therapy as: a socio-health profession that through the assessment of capacities and physical, mental, sensorial and social problems of the individual that aims, through proper treatment, to enable patients to achieve the greatest degree of independence as possible in their daily life, contributing to the recovery from illness or facilitating adaptation to disability [1]. The goals of occupational therapy on the users defined by ENOTHE (European Network of Occupational Therapy in Higher Education) [16] are: to maximize functional abilities, develop residual capacities, supply sunk functional deficits and achieve reintegration in society, living with a disability, but with the greatest degree of independence possible.

The occupational therapy process has different phases involving three types of actions: assessment, intervention, and analysis of results. So far, all occupational therapists adhere to a methodology in the process of assessment and intervention carried out in the centers they work [16]. These processes are virtually identical in all cases and determine the actions to be taken on each of the users and processes to be followed in their evaluation and subsequent rehabilitation. In the following paragraph, we are going to describe the process in a general manner:

(1) Initial intervention. It is a process guided by a semi-structured interview with the patient. During the interview be completed entirely manually and paper the following rating scales: Barthel Index, Lawton and Brody Scale, Scale Kels, Rating Scale Hobbies and Leisure (EVOTL), Leisure Interest List (LOII), Pfeiffer Questionnaire, Mini-Mental.

(2) Due to the complexity of the processes of initial assessment and reassessments, sometimes, it is carried out in different sessions, since they require a long period and the involvement of the person being evaluated.
(3) Evaluation report. The results of the assessments are compiled in a report, which is used as a basis for a treatment planning. After the intervention, a re-evaluation can be required if necessary.

Regarding applications for mobile devices used for evaluation and intervention in occupational therapy, there are currently no applications specifically designed for occupational therapy, although it is possible to find applications of sanitary type (used by people of various professional categories such as:. Doctors, physiotherapists, psychologists, nurses, etc.) that can be useful in tasks performed by the occupational therapist. We can classify these applications into two groups, depending on the device on which they run: wired devices and mobile devices. For wired devices, it is possible to find applications as PROIEC [19] that provides a structured intervention in the non-pharmacological field, intended for the person with dementia and rehabilitation of cognitive deficits particular material. For mobile devices, such as tablets or smartphones, some applications used in occupational therapy are: Baluh [4] that provides a communication solution for people who have difficulty speaking. It offers natural sounds, reading text, and is available in 14 languages and uses ARASAAC pictograms distributed under Creative Commons license. Baluh is used for intervention with children and adults with PDD (Pervasive Developmental Disorder). Picaa [11] is a support tool in the educational environment for users with special needs that facilitates and stimulates activities related to education. NeuroScores is a mobile application containing neurological scales, to evaluate stroke severity, progression, prognosis and outcome. It includes the rules for conducting a proper systematic neurological examination in the "Stroke Code". It calculates the score on the different scales of assessment, making a summary of the exploration and giving us a prognosis and treatment recommendation. Nerve Whiz is an application for medical professionals which shows the anatomy of the nerves. It studies the muscle innervation and areas of sensitivity loss making use of diagrams and charts. It is available for iOS devices. Glasgow scale is a mobile application that determines the level of consciousness using the Glasgow scale. It allows different user profiles depending on patient assessment. It is available for devices with Android.

To design the proposed approach, we evaluated different technologies. There are currently several operating systems for mobile devices that can be classified based on market share (data according to a study by International Data Corporation): Android (59%), iOS (23%), Symbian (6,8%), BlackBerry OS (6,4%), Bada (3,6%), Windows Phone. As can be seen, more than 80% of the sector is coped by the to the two leading mobile OS: Android and iOS. This prominence is due not only to the market data, but also the number of applications available, the maturity of the OS, the number of devices that possess, revenues, etc. One of the advantages of multiagent systems is the independence of operating system. To test the system we have developed a personal agent for the therapists that can be run on a iOS tablet.

3 MAS for Occupational Therapy

PANGEA is an agent platform to develop open multiagent systems [28], specifically those including organizational aspects such as virtual agent organizations. The platform allows the integral management of organizations and offers tools to the end user. Additionally, it includes a communication protocol based on the IRC standard, which facilitates implementation and remains robust even with a large number of connections. A general overview of the architecture is presented in Figure 1. The platform examines two modes of operation. In the first mode, the agents reside in the machine itself, while in the second mode the platform allows for the possibility of initiating all agents in different machines. The latter case has the disadvantage of allowing only minimal human intervention since it is necessary to previously specify the address of the machine where each of the agents are to reside; however it has the advantage of greater system distribution.

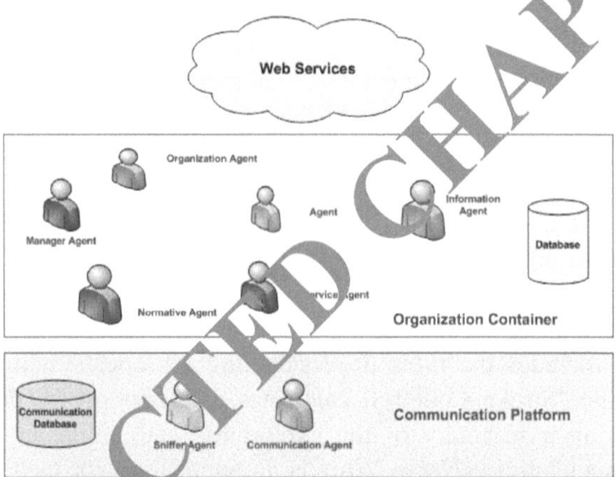

Fig. 1. PANGEA architecture

PANGEA has been used to develop a MAS for occupational therapy, where interface agents are installed on a iOS-based device and interact with the organization of agents. The use of PANGEA allow us to design the system in terms of organization, taking into account the human social infrastructures that participate in the occupational therapy problem. Thus, in the proposed solution we have defined a series of specialized agents:

- Therapist agent. This agent is installed in the mobile device (tablet) used by the therapist. The agent contains the valuation scales and the color cards that can be updated when connecting to the rest of the agents in the platform. The agent allows the therapist to write reports and to plan interventions.
- Store agent. This agent stores the information provided by the therapist agents related to patients' evaluation and intervention. Besides, this agent stores information about the different actors in the system (therapists and patients).

- Scales agent. This agent manages the different valuation scales that can be used by the therapist.
- Evaluator agent. This agent is specialized on analyzing the Information provided in the reports and obtain statistics about evaluations and interventions. The agent incorporates a case-based reasoning [11, 8, 9] system and defines rules to represent the knowledge obtained from the evaluations and interventions. In the future work this agent will incorporate a decision support system for the therapists to make recommendations based on past experiences.

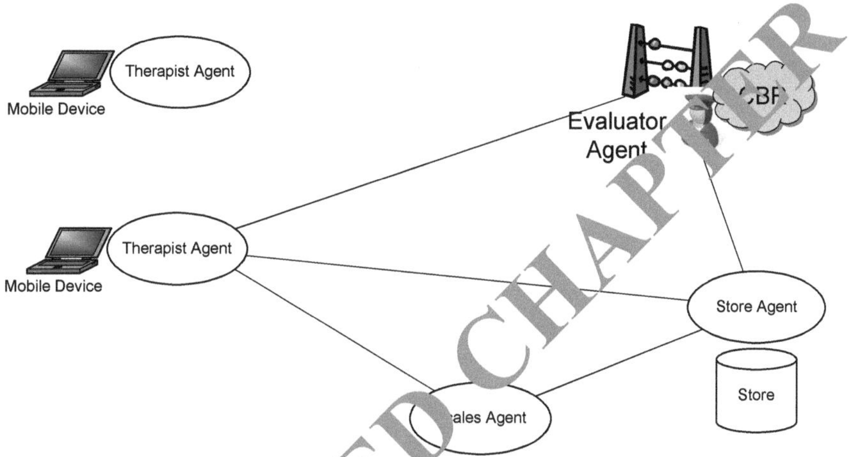

Fig. 2. Multi-agent architecture for occupational therapy

The proposed multi-agent system can be easily connected to similar MASs that would be installed in different health care scenarios [24, 26, 27], allowing interconnection and reuse of the resources available for occupational therapy. Besides, the use of the PANGEA architecture allow us to define organizational structures similar to those used in human societies. In this paper, we will focus on a particular multi-agent system without taking into account organizational aspects, and more specifically on the Therapist agent, which is installed on a mobile device and provides a tool to the therapist interact with the patient and to carry out the evaluations and interventions. The organizational aspects will be studied in further work.

4 Preliminary Results and Conclusions

After reviewing the state of the art, it has been found that the current therapeutic approach is mostly manual, and generates a lot of information that sometimes it is difficult to manage. Besides the use of a physical medium such as paper makes it difficult to find information, data mobility, and access to records. Moreover, in the intervention process, the use of color cards is repetitive and monotonous for the patient. Most of the times the cards are not up to date and the patient can only interact with them. In relation to the methodology followed by the occupational therapist, a new opportunity arises to improve the evaluation and intervention process. The process can be automated using a multi-agent system that can notably improve aspects such as computation and communication. In this sense, the use of agents installed in mobile devices facilitate the patient evaluation in a ubiquitous manner and an efficient storage and access to the required information. Furthermore, the use of the proposed multi-agent system provides a high flexibility for the management and monitoring of all the patients and therapeutic history. It also allows all therapists of a medical center (or even from different medical centers) to follow a common methodology, and access all the information in a ubiquitous manner. Regarding the patient, the use of an interactive device, enhances the experience in the intervention process, motivating them to perform therapeutic activities and facilitating access therapy processes. Figure 3 shows an example of a therapist agent. Figure 3.a shows a screenshot of the intervention phase, and more specifically some items used as color cards. Figure 3.b shows a screenshot of the evaluation phase. As can be seen in Figure 3.b, the therapist can choose different valuable scales and write a final report for each of the patients.

Fig. 3. Screenshot of the therapist agent. a) Flash cards. b) valuation scales

The proposed approach requires a mobile medium size, given the characteristics of the contents thereof. It must be portable but of an appropriate size to display forms and images to a readable size for patients and therapists. It must allow access to the internet and be friendly to both the application for single-strength SSOO own interface. In this case we opt for a tablet type device to install the therapist agent. The use of the PANGEA platform allows an easy remote connection of the therapist agent.

The proposed approach was evaluated in a medical center in the north of Spain with 10 users and the initial results were promising. In particular, the system was tested with 10 patients attending a pre-laborer workshop for people with mental and intellectual disabilities. The sessions consisted on conducting an initial evaluation, recognition of numbers and letters, colors, coins and banknotes, the same way the traditional material was used to support the acquisition of skills and abilities. From the point of view of a professional of neural rehabilitation, this application provides a substantial improvement in both the occupational therapist and the patients attending the medical center. One of the advantage mentioned by the medical staff is the mobility and independence of local resources. Besides, the therapist highly valuated the statistical analysis provided by the system, which notably improves the previous manual processes. Another important point for the therapist is the quality of the cards (images).

Our future work will focus on the evaluation of the multi-agent system and inclusion of organizational aspects. We will also focus on a more general group of individuals to evaluate the proposed approach.

Acknowledgments. Work partially supported by the Spanish Government through the project iHAS (grant TIN2012-36586-C03).

References

1. APETO (last accessed 2014), `http://www.apeto.com`
2. Bajo, J., Corchado, J.M.: Evaluation and monitoring of the air-sea interaction using a CBR-agents approach. In: Muñoz-Ávila, H., Ricci, F. (eds.) ICCBR 2005. LNCS (LNAI), vol. 3620, pp. 50–62. Springer, Heidelberg (2005)
3. Bajo, J., De Paz, J.F., Rodríguez, S., González, A.: Multi-agent system to monitor oceanic environments. Integrated Computer-Aided Engineering 17(2), 131–144 (2010)
4. Baluh (last accessed 2014), `http://blog.baluh.org/`
5. Borrajo, M.L., Baruque, B., Corchado, E., Bajo, J., Corchado, J.: Hybrid neural intelligent system to predict business failure in small-to-medium-size enterprises. International Journal of Neural Systems 21(04), 277–296 (2011)
6. Carneiro, D., Castillo, J.C., Novais, P., Fernández-Caballero, A., Neves, J.: Multimodal Behavioural Analysis for Non-invasive Stress Detection. In: Expert Systems With Applications, vol. 39(18), pp. 13376–13389. Elsevier (2012) ISSN: 0957-4174
7. Corchado, J.M., Aiken, J.: Hybrid artificial intelligence methods in oceanographic forecast models. IEEE Transactions on Systems, Man, and Cybernetics, Part C: Applications and Reviews 32(4), 307–313 (2002)
8. Corchado, J.M., De Paz, J.F., Rodríguez, S., Bajo, J.: Model of experts for decision support in the diagnosis of leukemia patients. Artificial Intelligence in Medicine 46(3), 179–200 (2009)
9. Corchado, J.M., Bajo, J., De Paz, J.F., Rodríguez, S.: An execution time neural-CBR guidance assistant. Neurocomputing 72(13), 2743–2753 (2009); De Paz, J.F., Rodríguez, S., Bajo, J., Corchado, J.M.: Case-based reasoning as a decision support system for cancer diagnosis: A case study. International Journal of Hybrid Intelligent Systems 6(2), 97–110 (2009)
10. Creek, J., Lougher, L.: Occupational Therapy and Mental Health. Churcill Livingston - Elsevier (2008)
11. Fernández López, A., Rodríguez Fórtiz, M.J.: Dispositivos móviles iPod touch y iPad para Aprendizaje en Educación Especia. In: Arnaiz, P., Hurtado, M.D., Soto, F.J. (Coords.) 25 Años de Integración Escolar en España: Tecnología e Inclusión en el ámbito educativo, laboral y comunitario, Consejería de Educación, Formación y Empleo, Murcia (2010)

12. Fdez-Riverola, F., Corchado, J. C.: based system for forecasting red tides. Knowledge-Based Systems 16(5), 321–328 (2003)
13. Fraile, J.A., Bajo, J., Corchado, J.M., Abraham, A.: Applying wearable solutions in dependent environments. IEEE Transactions on Information Technology in Biomedicine 14(6), 1459–1467 (2011)
14. Garijo, F., Gómez-Sanz, J.J., Pavón, J., Massonet, P.: Multi-agent system organization: An engineering perspective. In: Pre-Proceeding of the 10th European Workshop on Modeling Autonomous Agents in a Multi-Agent World, MAAMAW 2001 (2001)
15. Griol, D., García-Herrero, J., Molina, J.M.: Combining heterogeneous inputs for the development of adaptive and multimodal interaction systems, Advances in Distributed Computing And Artificial Intelligence Journal
16. Hagedorn, R.: Occupational Therapy Perspective and processes. Churchill Livingstone - Elsevier (1995)
17. Pavon, J., Sansores, C., Gomez-Sanz, J.J.: Modelling and simulation of social systems with INGENIAS. International Journal of Agent-Oriented Software Engineering 2(2), 196–221 (2008)
18. Pinzón, C.I., Bajo, J., De Paz, J.F., Corchado, J. S.-M.: An adaptive hierarchical distributed multi-agent architecture for blocking malicious SOAP messages within Web Services environments. Expert Systems with Applications 38(5), 5486–5499
19. PROIEC (last accessed 2014), `http://www.terapia-ocupacional.com/ayudas-tecnicas/IVE/proiec.shtml`
20. Rodríguez, S., Pérez-Lancho, B., De Paz, J.F., Bajo, J., Corchado, J.M.: Ovamah: Multiagent-based adaptive virtual organizations. In: 12th International Conference on Information Fusion, FUSION 2009, pp. 990–997 (2009)
21. Rodríguez, S., de Paz, Y., Bajo, J., Corchado, J.: Social-based planning model for multiagent systems. Expert Systems with Applications 38(10), 13005–13023 (2011)
22. Schuler, D.: Social computing. Communications of the ACM 37(1), 28–29 (1994)
23. Sánchez-Pi, N., Carbó, J., Molina, J.M.: A Knowledge-Based System Approach for a Context-Aware System. Knowledge-Based Systems 27, 1–17 (2012)
24. Tapia, D.I., Alonso, R.S., De Paz, J.F., Corchado, J.: Introducing a distributed architecture for heterogeneous wireless sensor networks. In: Omatu, S., Rocha, M.P., Bravo, J., Fernández, F., Corchado, E., Bustillo, A., Corchado, J.M. (eds.) IWANN 2009, Part II. LNCS, vol. 5518, pp. 116–123. Springer, Heidelberg (2009)
25. Tapia, D.I., Abraham, A., Corchado, J.M., Alonso, R.: Agents and ambient intelligence: case studies. Journal of Ambient Intelligence and Humanized Computing 1(2), 85–93 (2010)
26. Tapia, D.I., Rodríguez, S., Bajo, J., Corchado, J.M.: FUSION@, a SOA-based multi-agent architecture. In: International Symposium on Distributed Computing and Artificial Intelligence (DCAI 2008), pp. 99–107 (2008)
27. Tapia, D.I., Alonso, R.S., García, Ó., de la Prieta, F., Pérez-Lancho, B.: Cloud-IO: Cloud Computing Platform for the Fast Deployment of Services over Wireless Sensor Networks. In: 7th International Conference on Knowledge Management in Organizations: Service and Cloud Computing, pp. 493–504 (2013)
28. Zato, C., Villarrubia, G., Sánchez, A.: PANGEA - Platform for Automatic Construction of Organizations of Intelligent Agents. In: Distributed 9th International Conference Computing and Artificial Intelligence (2012)

A Gaussian Particle Swarm Optimization for Training a Feed Forward Neural Network

Haydee Melo and Junzo Watada

[1] Haydee Melo at Graduate School of Information, Production and System, Waseda University, Fukuoka, Japan
melo.haydee@asagi.waseda.jp
[2] Junzo Watada at Graduate School of Information, Production and Systems, Waseda University, Fukuoka, Japan
junzow@osb.att.ne.jp

Abstract. This paper proposes a Gaussian-PSO algorithm which provides the optimized parameters for Feed Forward Neural Network. Recently the Feed Forward Neural Network is widely used in various applications as a result of its advantages such as learning capability, auto-organization and auto-adaptation. However the Neural Network has the disadvantage itself to slowly converge and get easily trapped in a local minima. In this paper, Gaussian distributed random variables are used in the PSO algorithm to enhance its performance and train the weights and bias in the Neural Network. In comparison with the Back Propagation Neural Network, the Gaussian PSO-Neural Network faster converges and is immuned to the local minima.

1 Introduction

Recently, Neural Networks are most widely used in non-linear applications such as regression analysis, time series prediction, modeling, classification, image processing and pattern recognition. The most widely used algorithm for training this network is a back-propagation (BP) algorithm. Many learning techniques has been developed for the BP as gradient descent [1], resilient, BGFS quasi-Newton, one-step secant, Levenberg-Marquad [4] and Bayesian regularization whose objective is to minimize the error. The BP algorithm has the problem to get trapped in local minima at complex problems, also its convergence speed is slow [3]. Researchers made some improvements on the convergence of the BP in the Neural Network, however these improvements involve the change of the function for excitation and without ideal results [7]. In the current research different types of algorithms have been used for training the Neural Network such as genetic algorithm (GA), simulating annealing algorithm (SAA), and Particle Swarm Optimization algorithm (PSO) [8][10]. PSO is a population-based algorithm to find optimum solutions by using a swarm like algorithm where particles collaborate and share information among them. In this paper an improved PSO is used with Gaussian random variables to straighten out the convergence and avoid local minima.

J.B. Perez et al. (eds.), *Trends in Practical Applications of Heterogeneous Multi-agent Systems. The PAAMS Collection*, Advances in Intelligent Systems and Computing 293,
DOI: 10.1007/978-3-319-07476-4_8, © Springer International Publishing Switzerland 2014

2 Particle Swarm Optimization

PSO was originally proposed by Kennedy and Eberhart [5] in 1995, by mimicking the collective learning of individuals when they are in a group.

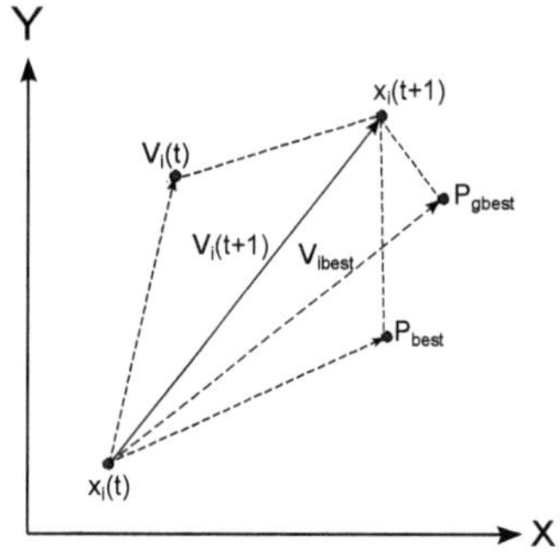

Fig. 1. Classic PSO

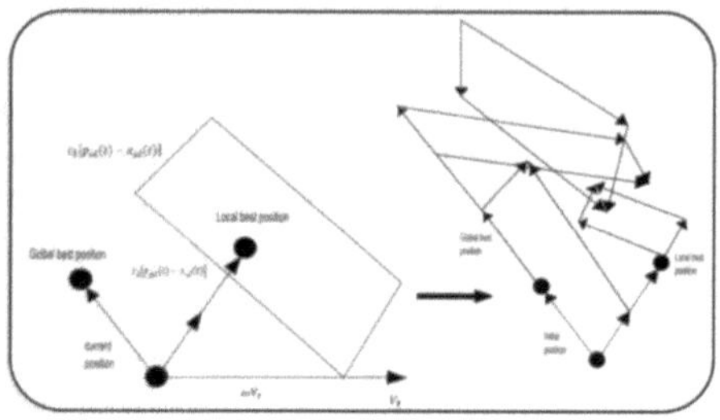

Fig. 2. Visualization of the Swarm

In PSO each particle in the flock is initialized in uniform randomness over a searching space with a dimension N. At each iteration, each particle updates velocity and position based on its best position P_b and the neighbors' best position P_g. The number of particles n, the position of the ith particle is expressed as $X_i = (x_{i1}, x_{i2}, \cdots, x_{iN})$; the best position of the ith particle is expressed as $P_{ib} = (P_{i1}, P_{i2}, \cdots P_{iN})$, the best position of all the swarm is $P_g = (P_{g1}, P_{g2}, \cdots, P_{gN})$, the velocity of the ith particle is $V_i = (v_{i1}, v_{i2}, \cdots, v_{iN})$. Then the ith particle velocity and position at $(t+1)$can be calculated by using them.

Later Eberhart introduced weights into the original PSO. These weights changed adaptively to balance the global search and the local search. This inertial weight w is used in the algorithm in order to refine the search around the global optimum. The APSO is formulated as follows:

$$v_{id}(t+1) = w * v_{id}(t) + c_1 * r_1[p_{id}(t) - x_{id}(t)] + c_2 * r_2 * [p_{gd}(t) - x_{id}(t)]$$
$$x_{id}(t+1) = x_{id}(t) + v_{id}(t+1)$$
$$1 \le i \le n1 \le d \le D \quad (1)$$

where w is the inertial weight. In the beginning of iterations the weight is decreased rapidly, after a certain number of iterations the weight is decreased slowly by using the following formula:

$$w = \begin{cases} w_0 - ((w_0 - w_1)/g_1) * g & 1 \le g \le g_1 \\ (w_1) * e^{g_1 - g)/k}, & g_1 \le g \le g_2 \end{cases} \quad (2)$$

where w_0 is the initial inertial weight and w_1 is the inertial weight of the linear reduction strategy and g_2 is the total number of generations in the algorithm, g_1 is the number of generations for the linear reduction strategy which is decided empirically, and g is the actual generation number in the range $[1, g_2]$. In the reduction weight strategy, k is used to adjust the non-linear reduction, the different values in k modify the curve [10]. In APSO, if the initial parameters are not properly set, the convergence speed becomes slow in the search near to the global optimum. In order to avoid this, Gaussian random variables instead of uniform random variables are used to escape from the local minima.

3 Gaussian Particle Swarm Optimization

3.1 Gaussian Random Variables

Gaussian probability distribution improves the local convergence of the PSO, by balancing the trade-off between exploitation and exploration in the velocity update. If we analyze the probability distribution of a particle without initial velocity, the position of the particle is considered to be inside of a parallelogram (uniform probability). Then the particle's actual position is considered as the vertex of the next parallelogram where the new position can be found. After many interactions we observe that the parallelograms overlap as shown in Fig. 2. The center of the parallelograms is observed to be in the opposite position of the initial position. By using the central limit theorem we assume an Gaussian distribution. Therefore, the Gaussian distributed random variables $G(\mu, \sigma^2)$ are introduced instead of the uniformly distributed random variables:

$$\begin{aligned} v_{id}(t+1) = v_{id}(t) + c_1 * G(\mu, \sigma^2)[p_{id}(t) - x_{id}(t)] \\ + c_2 * G(\mu, \sigma^2) * [p_{gd}(t) - x_{id}(t)] \\ x_{id}(t+1) = x_{id}(t) + v_{id}(t+1) \\ 1 \leq i \leq n1 \leq d \leq D \end{aligned} \quad (3)$$

3.2 Ziggurat Algorithm

The Ziggurrat algorithm is a method used for sampling Gaussian random variables, based on a rejection sampling algorithm where a distribution to be slightly larger than the desired distribution is generated as depicted in Fig. 3 [6].

The Ziggurat algorithm generates a point x from a probability distribution which is made up of n-equal-area regions; $n - 1$ rectangles can cover the desired distribution, on top of a non-rectangular base that includes the tail of the distribution.Then the right hand edge of each rectangle is placed to cover the distribution, as depicted in Fig 3 as the gray area in the rectangle. Therefore, some of the area in the top right of the rectangle is outside of the distribution (points with $y > f(x)$). However R_0 is an exception and is entirely within the probability distribution function (pdf).Then the $v(r)$ area of each rectangle can be calculated.

4 Feed Froward Neural Network

A Feed Forward Neural Networ (FFNN) is principally formed of different layers: input layer, hidden layer and output layer as depicted in Fig. 4. All the layers are connected

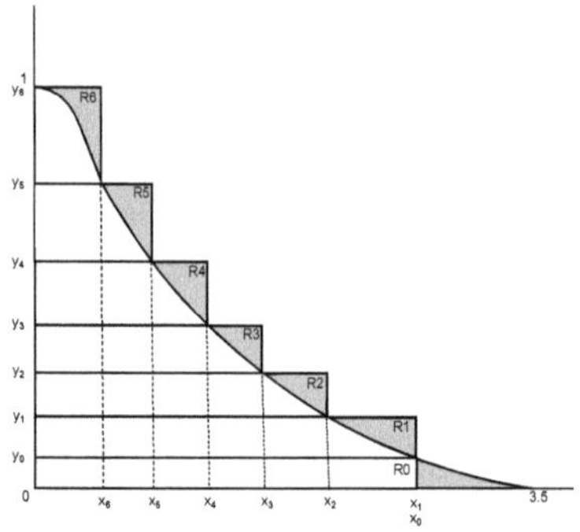

Fig. 3. The Zigurrat algorithm with 7 rectangle

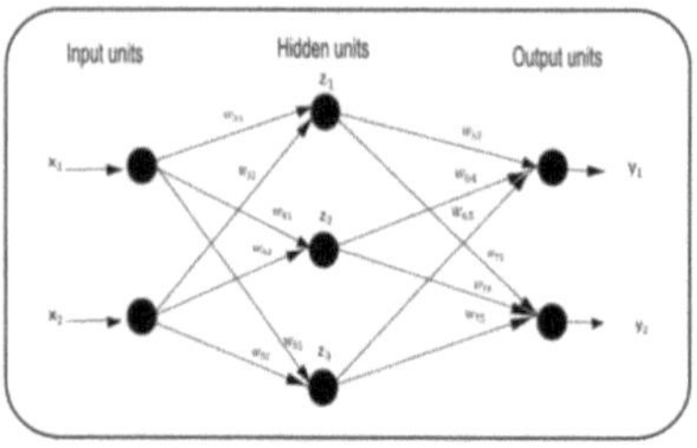

Fig. 4. Neural Network Structure

between them by links that are called weights. The learning in the NN is performed by determining the optimized values in the weights that connect each layer. Suppose that the input layer had n units (or neurons), the hidden layer has H hidden units, and the output layers has O output units. Assuming that the activation function is a commonly used sigmoid function. Each input unit $(X_i, i = 1, ..., n)$ receives an input signal x_i and transfers this signal to all the units in the next layer (hidden layer), each hidden unit $(Z_j, j = 1, ..., H)$ sums its weighted input signals as:

$$z_{inj} = \sum_{i=0}^{n} x_i w_{ij} - \theta_{0j} \tag{4}$$

where w_{ij} is the weight from the ith unit of the input layer to the jth unit of the hidden layer, θ_{0j} denotes the bias on the hidden unit j; x_i is the ith input unit. z_{inj} is the weight input sum in the hidden layer. The activation function is computed:

$$z_j = \frac{1}{1 + exp(-(z_{inj}))} \tag{5}$$

Each output unit $(Y_k, k = 1, ...O)$ sums its weighted input signals where w_{jk} is the weight from the jth hidden unit to the kth output unit, θ_k is the threshold of the kth output unit.

$$y_{ink} = \sum_{j=0}^{H} z_j w_{jk} - \theta_k \tag{6}$$

and applies its activation function to compute its output signal,

$$y_k = \frac{1}{1 + exp(-(y_{ink}))} \tag{7}$$

Each output unit $(Y_k, k = 1, ...O)$ receives the target pattern and the error E can be calculated as:.

$$E = \frac{1}{O} \sum_{i=0}^{O} (y_i - Y_i)^2 \tag{8}$$

where $(Y_i - y_i)$ is the error of the actual output and the desired output of the ith output. Then the fitness function for the PSO can be formulated as

$$fitness(X_i) = E(X_i) \tag{9}$$

5 Encoding Strategy and Parameter Selection

In this paper a matrix strategy for encoding the weights because of the advantage of easy understand of the weights structure in the network. For example using a 2-3-2 network structure as despited on Fig. 4, the matrix can be written as follows: the first column is constituted by w_{31}, w_{41}, w_{51}, there are the weights from input unit x_1 to the hidden layer, then second column are the weights from input unit x_2 to the hidden units w_{32}, w_{42}, w_{52}. The next columns correspond to the weights from hidden units to the output units. In order to simplify the encoding one matrix configuration is used, so w_{63}, w_{64}, w_{65} are weights from the hidden units to the output unit y_1, and w_{73}, w_{74}, w_{75} are weights from the hidden units to the output unit y_2. The experiment is carried out with $c_1 = 1.2$ and $c_2 = 1.2$ that achieved the best results and the range for v_{min} and v_{max} is [-10,10]. Then the search space range of (-100,100) is observed to be the best range with a swarm size of 25 particles.

6 Gaussian PSO-Feed Forward Neural Networks

The PSO is a population based algorithm, but has disadvantage that the search around the global optima is slow. To mitigate this disadvantage, Gaussian random variables are adopted. The idea is to use the advantages of the Gaussian-PSO to search for the optimum weight values of the FFNN. The procedure for this GPSO-NN can be summarized as follows:

- Step 1: Initialize the positions and velocities and iteration t=1.
- Step 2: Introduce the weight values in the FFNN to evaluate each initialized particle‘s fitness value.
- Step 3: If the maximal iterative generations is reached, go to Step 6, otherwise continue to Step 4
- Step 4: The best particle of the current particle is stored. The positions and velocities of all the particles are updated according to Eqs.(2) and (3) for position and velocities, if a new particle flies beyond the boundary $[x_{min}, x_{max}]$, the new position will be set to x_{min} or x_{max} ; if a new velocity is beyond the boundary $[v_{min}, v_{max}]$, the new velocity will be set to v_{min} or v_{max}.
- Step 5: Evaluate each new particle‘s fitness value, and the worst particle is replaced by the stored best particle. If the ith, particle‘s new position is better than P_b, P_g is set to the new position of the ith particle. If the best position of all the new particles is better than P_g , then P_g is updated.
- Step 6: Output the global optimum

7 Simulation Results

For initial values, every weight in the network was initially set in a range of [-50,50], and all the thresholds in the network were 0. The initial inertial weight was set to 1.8. The initial population was set to 100. For testing the proposed method a 1-S-1 Network structure was used for approximating the function $f = sin(2x)e^{-x}$. The training algorithms for this structure were the BP algorithm and the Gaussian-PSO. In the structure 1-S-1, where $S = 3, 5, 7$, and the x_i was obtained from the range $[0, \pi]$, and the sampling interval was 0.03 for the training set and 0.1 for the testing set in the range $[0.02, \pi]$.For different hidden units in the hidden layer, the test was run three times. The

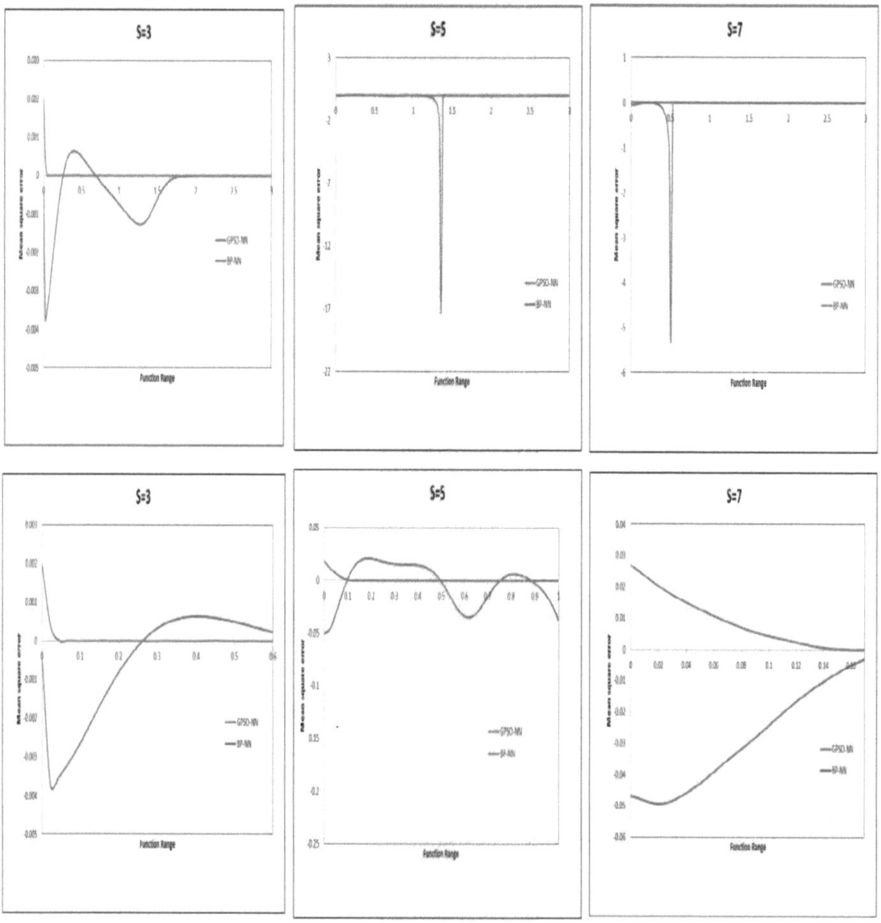

Fig. 5. The training error curves of the testing samples for the function $f = sin(2x)e^{-x}$ for the two training algorithm: Gaussian-PSO (blue line) and Back Propagation algorithm (red line) respectively for different S.

maximum iteration for the BP algorithm was 2000 times and for the Gaussian-PSO a maximum was set to 1500 iterations. Table 1 gives the performance comparison for the three algorithms. From Table 1 we can compare the CPU used in each algorithm and the accuracy for both training methods.

Table 1. Comparison of the performances of the GPSO and the BP for the function approximation problem $f = sin(2x)e^{-x}$

Hidden unit	GPSO		BP	
s1	MSE	Time(s)	MSE	Time(s)
3	5.67e-005	3.35	2.38e-003	9.33
5	1.27e-009	3.76	1.22e-003	8.16
7	1.81e-007	4.32	8.15e-005	9.35

Also we conclude in Table 1, that the selection of different hidden unit number has an influence in the performance of both training methods. For the approximation function $f = sin(2x)e^{-x}$

8 Conclusions

A Gaussian Particle Swarm optimization for training a FFNN for function approximations is presented in this paper. The objective is to use the strong ability of the PSO in the global search and enhance its deficiency in the local search by introducing Gaussian random variables into the updates equation in the algorithm. In this hybrid algorithm a weight strategy is also introduced in the Gaussian-PSO that reduces the weight rapidly in the first generation and then slowly when it is near to the global optima. For testing the hybrid algorithm a function approximation problem for a non-linear function was used. The experiment was realized with different number of hidden units in the Neural Network configuration for analyzing the performance of BP algorithm and the Gaussian-PSO. The performance of both algorithms is improved when the number of hidden units is increased. However, we observed that the number of computations that Gaussian-PSO required for learning is fewer than the back-propagation algorithm. For future work, we will apply this algorithm for a real application and to incorporated this training algorithm into recurrent neural networks and recurrent fuzzy network.

References

1. Chen, C.H., Lai, H.: An empirical study of the Gradient descent and the conjugate gradient backpropagation neural networks. In: Proceedings OCEANS 1992 Mastering the Oceans Through Technology, vol. 1, pp. 132–135 (1991)
2. Eberhart, R.C., Shi, Y.: A Modified Particle Swarm Optimizer. In: Proceedings of IEEE International Conference On Evolutionary Computation, pp. 69–73 (1998)

3. Gudise, V.G., Venaayagamoorthy, G.K.: Comparison of Particle Swarm Optimization and Back Propagation as Training algorithms for Neural Networks. In: Proceedings of the IEEE Swarm Intelligence Symposium 2003 (SIS 2003), pp. 110–117 (2003)
4. Hagan, M.T., Menhaj, M.B.: Training feed forward networks with the Marquadt algorithm. In: IEEE International Conference on Tools with Artificial Intelligence, pp. 45–49 (1997)
5. Kennedy, J., Eberhart, R.C.: Particle Swarm Optimization. In: Proc. IEEE Conf. Neural Networks IV, vol. 4, pp. 1942–1948 (1995)
6. Marsaglia, G., Wai, W.T.: The Ziggurat Method for generating Random Variables. Journal of Statistical Software 5(8), 1–7 (2000)
7. Moller, M.F.: A scaled conjugate gradient algorithm for fast supervised learning PB-339 Reprint, Computer science Department, University of Aaurhus, Denmark (1990)
8. Salerno, J.: Using the particle swarm optimization technique to train a recurrent neural network. In: Proc. of Ninth IEEE Int. Conf. on Tools with Artificial Intelligence, pp. 45–49 (1997)
9. Settles, M., Raylander, B.: Neural network learning using particle swarm optimizers. In: Advances in Information Science and Soft Computing, pp. 224–226 (2002)
10. Zhang, J.: A hybrid particle swarm optimization-back propagation algorithm for feed forward neural network training. Applied Mathematics and Computation 185, 1026–1037 (2007)

Cluster Analysis of Patients Suffering from Addictions

Catherine Combes

University of Lyon - F-42023 Saint-Etienne
UMR CNRS 5516 - HUBERT CURIEN Laboratory
Jean Monnet University
18 rue Benoît Lauras
42023 Saint-Etienne cedex 2- France
catherine.combes@univ-st-etienne.fr

Abstract. We investigate the contribution of unsupervised learning to identify patient's profiles suffering from addictions. We propose a new clustering approach based on coupling b-coloring of graph and Bregman hard clustering algorithm in order to automatically find the number of categories or groups of patients and the "best" representative patients' profile of each group. The study was carried out in close collaboration with the French co-operative health organization called the "Centre Mutualiste d'Addictologie", an aftercare centre for addictions. The quantitative data arises from a cohort of seven different aftercare centres for addiction located in France. The study concerns 301 patients suffering from dependence (addictions with psychoactive substances and/or behaviour addictions).

1 Introduction

An addiction is baffling and is mostly characterized by a loss of control (compulsive behaviour) in limiting intake and increased tolerance (a higher dose is required to achieve the same effect). Many faces of addiction can be identified: addictions with psychoactive substances and behavioural addictions. Behavioural addictions (also called process addiction or non-substance-related addiction) concern food, sex, computers, video games, internet, gambling, work, shopping... Several health and social problems are associated with use of and dependence on alcohol, tobacco and illicit substances and/or non-substance-related addiction. Concerning addiction pathologies, a wide variety of diseases may occur: somatic complications and/or mental illness and multiple factors influence the addiction acuity. The aim of this paper is to find features-patterns that relate the patients suffering from addiction using cluster analysis. Data clustering is a fundamental "unsupervised" learning procedure that has been extensively studied across different disciplines over several decades allowing to assign a set of objects into groups (called clusters) so that the objects in the same cluster are similar (in some sense) to each other and the different groups are dissimilar. Cluster analysis is meaningful, useful, or both and is a useful starting point for other purposes, such as data summarization. From 301 patients concerning 7 aftercare centres for addiction, we observe 4 factors: *Addiction*, *Somatic*, *Psychiatric* and *Social*.

J.B. Perez et al. (eds.), *Trends in Practical Applications of Heterogeneous Multi-agent Systems. The PAAMS Collection*, Advances in Intelligent Systems and Computing 293,
DOI: 10.1007/978-3-319-07476-4_9, © Springer International Publishing Switzerland 2014

In fact, patient's evaluation regarding the 4 factors can been see as the set of words to learn in order to extract knowledge (data summarization).

The proposed clustering approach is based on coupling graph theory (b-coloring of a graph corresponding to a special case of graph labelling) in order to identify the number of clusters and Bregman hard clustering algorithm.

The core of this article is made up of five sections. In section 2, we briefly discuss some related works. Section 3 describes our proposed clustering approach. In section 4 we present our experiments concerning population-based sample suffering from addiction. The validation of the clusters is discussed using three indices of cluster validity (Dunn's separation index, Davies-Bouldin's index and silhouette validation method). A clinical validation made by the resident doctor (psychiatrist) of the after-care centre for addictions is described in order to validate the features of the clusters. Conclusions and future research are discussed in section 5.

2 Related Works

Graph theory has long been applied in cluster analysis but essentially in hierarchical clustering. Some works use graph theory for hard partition clustering. From dissimilarities among data items, the entire data items can be represented as a graph structure, where each data item is mapped to a vertex and the vertices are connected by edges with the corresponding dissimilarities. Recently, [Irving, Manlove, 1999] proposed the notion of b-coloring of undirected graphs. Based on this, [Elghazel et al., 2006] proposed a clustering method, but it did not consider the quality of clusters. A re-coloring algorithm was proposed to reflect the quality of clusters [Elghazel et al., 2008]. It was still restrictive in terms of the explored search space due to its greedy and sequential process. [Ogino, Yoshida, 2010] proposed a re-coloring algorithm to do non-greedy search for the admissible colors of vertices. But cluster validity indices are used in order to determine the "optimal" partition and the literature shows the limits of these indices. We also note that the performance of re-coloring based on precision cannot be utilized as the quality measure in any algorithms directly, since it is calculated based on the "true" labels and "true" labels are unavailable for clustering or in unsupervised learning in general.

And other problem rarely mentioned is the impact of distance measure. Many approximation algorithms and techniques for this minimization problem have been developed when the dissimilarity measure used is a metric such as the Euclidean distance (known as the Euclidean k-median problem), or when the squared Euclidean distance is used (known as the Euclidean k-means problem). However, until recently, for non-metric dissimilarity measures almost no approximation algorithms were known. This stands in sharp contrast to the fact that many non-metric dissimilarity measures are used in various applications, such as Mahalanobis, Itakura-Saito and Kullback-Leibler divergences. These instances correspond to a broader class of dissimilarity measures that has only recently attained considerable attention: the class of Bregman divergences. It is based on information-theoretic criteria have been widely utilized in statistical analysis, machine learning, and data mining. Bergman hard clustering [Banerjee, 2005] corresponds to a quantization problem that involves minimizing the loss in Bregman information and it is very similar to squared distance based iterative k-means – except that distortion function is general class of Bregman Divergence. For categorical, data clustering, the relationship between entropy-based criterion and other criteria (including Minimum Description Length (MDL), Kullback–Leibler (KL) divergence, etc.) is shown in [Li et al., 2004].

In order to take account the limit of these works, we propose a methodology consisting of two successive procedures:

1. Identification of the number of clusters using:
   - b-coloring of the threshold graph,
   - searching the optimum dominating color for each vertex,
2. Improving the clustering results using entropy as optimization criteria with Bregman hard clustering algorithm.

3 Clustering Analysis Methodology

Graph coloring is a special case of graph labelling [Distel 2006]. The proper b-coloring problem [Irving, Manlove, 1999] is the assignment of colors to the vertices of one graph with two conditions: (i) Adjacent vertices have different colors, (ii) For each color, there exists at least one vertex having these neighboring vertices in all other colors. This vertex is called dominating vertex.

From the dissimilarity or distance matrix, the complete graph $G=(V,E,w)$ is defined by:

- ✓ a finite set $V=\{v_1,...,v_n\}$ of nodes $(/V/=n,\ v_i$ correspond to the observation o_i and n is the number of observations),
- ✓ a finite set $E=\{e_1,..., e_m\}$ of arcs over $V\times V$ with $(/E/=m)$,
- ✓ a function $w\colon E \rightarrow \mathbb{R}$ with $w(v_i;v_j) = D(o_i,o_j)$ where $D(.,.)$ is the dissimilarity or distance measure between the two observations. We choose the Generalized Measure Distance (GDM) proposed by Walesiak (Walesiak, 2002 pp. 33-43) because this measure can be used for data measured on ratio, interval, and ordinal scale.

Starting from G, the partial threshold graph $\boldsymbol{G}_{\geq s}$ corresponds to the superior threshold graph associated where threshold value s is $max_i\ min_j\ \{d_{ij}= D(o_i, o_j)/\ i\neq j\}$ (from dissimilarity or distance matrix between the observations).

We remove all edges with a weight (value of d_{ij}) larger or equal than the threshold s. Consequently, after removing these edges, we obtain a *filtered graph* called *threshold graph* $G_{\geq s}\ (V;E_{\geq s};w)$, such that $E_{\geq s} = \{(v_i;v_j)\ |\ w(v_i;\ v_j) = d_{ij}\ et\ \ w(v_i;\ v_j) \geq s\}$. We verify if this *threshold graph* $G_{\geq s}$ is connected.

The proper coloring of the graph $G_{\geq s}\ (V;E_{\geq s};w)$ consists to assign one color to all nodes (vertices) such that two adjacent nodes do not have the same color.

The coloring is called *b-coloring*, if for each color $\boldsymbol{C_k}$, it exists at least one vertex v_i colored $\boldsymbol{C_k}$ having the vertices in its neighborhood of different colors. The vertex v_i is called dominating vertex for the color $\boldsymbol{C_k}$.

The proposed algorithm aims to obtain a partition of the filtered graph $G_{\geq s}$ into disjoined color classes $\{C_1;C_2;\ ...;C_k\}$ that represent clusters. It is composed of two parts.

The first step of the algorithm constructs a proper coloring of the *threshold graph* $G_{\geq s}$. The second step consists in the re-coloring of the obtained graph using *b-coloring* of graph.

First some definitions are given, which are used in the procedure "*b-coloring algorithm pseudo code*" described below-written:

- Let Δ be the maximum degree of $G_{\geq s}$ obtained by the function *degree*(v_i) which allows calculating the degree of the vertex $v_i \in V$.
- Let $c(v)$ be the color of the vertex v in the graph $G_{\geq s}$.
- For every v, $N(v)$ is the set of adjacent vertices to v.
- The set of colors of $N(v)$ is $Nc(v)$.
- We note L the color set used in the graph $G_{\geq s}$ with $L=\{1,2,\ldots, \Delta+1\}$ corresponding to the color numbers.
- Let Dc be the set of dominating colors. A dominating color corresponds to the color of at least one dominating vertex. Initially $Dc=\varnothing$.
- Let T be an ordered set of vertices sorted by decreasing degree. Initially $T=\varnothing$.

When $L = Dc$, $|L|$ gives the number of dominating colors and hence the number of clusters and it is possible to identify the representative of each dominating color (which corresponds to cluster medoids). However it is necessary to improve this approach, because b-coloring based clustering algorithm did not optimize the quality of clusters. So we propose to use an approach based on Bregman hard clustering (Banerjee *et al.*, 2005) and the number of clusters and the corresponding representative object of each cluster are given in input of the algorithm.
The principle of the proposed algorithm is the following.

1. Select the first $v \in V$ such that $degree(v) = \Delta$
2. Assign the color 1 to $c(v)$
3. Add v to T
4. **foreach** $v_i \in N(v)$ (neighborhood of v) **do**
   4a. Assign a color to v_i such that already colored adjacent vertices have different colors
   4b. **if** more than one color is available **then**
      Assign (among the available colors) to v_i the color h of the vertex having one of available colors such that the distance/dissimilarity of this vertex to the vertex v_i is minimum
   4c. Add v_i to T
5. Remove v to T
6. Repeat steps 4 - 5 with v which is the first element of T, until T is empty (all vertices are colored)
7. Search if there exit dominating vertices (a dominating vertex v_i have this neighborhood in all the other colors)
8. Add to Dc the corresponding colors of the dominating vertices
9. Select color number $k \in L \backslash Dc$
10. **foreach** vertices v_i having the color k ($c(v_i)=k$) **do**
   10a. Search the available colors not in the neighborhood of v_i | $L \backslash Nc(v_i) \cup \{k\}$
   10b. **if** more than one color is available **then**
      Assign (among the available colors) to v_i the color h such that the average distances (or dissimilarities) of all vertices having the same color h is minimized
11. Remove k to L
12. Update Dc: Search if dominating vertices
13. Repeat steps 9 to 12 until $L=Dc$

4 Experiments: Patients Suffering from Addictions

For each of the 4 factors (Addiction, Somatic, Psychiatric and Social), seven aspects are evaluated. In the article, we only describe addiction and psychiatric factors. The aim is to identify psychiatric co-morbidity regarding addiction criteria concerning 301 patients suffering from addictions and treated in various addiction centres in France. Each variable can be evaluated using the three items value: "no", "yes with normal acuity", "yes with acuity ++".

Many different indices of cluster validity have been proposed in order to measure goodness of a clustering. We retained three of them:

- Dunn's separation index [Dunn, 1974],
- Davies-Bouldin's index [Davies, Bouldin, 1979],
- Silhouette Validation Method [Rousseeuw, 1987].

In the two following sections, we present the obtained results: the characteristics of the clusters and the corresponding indices of cluster validity.

4.1 Addictions

In addiction factor, the corresponding variables are:

Var 1: Alcohol consumption ≥ 14 IDU (International Units of Drink),
Var 2: Number of coming off drugs greater than 2,
Var 3: Tobacco dependence,
Var 4: Cannabis consumption,
Var 5: Substitute opiate prescribing,
Var 6: Other drug dependence,
Var 7: Behavioural addiction without eating disorders.

We identify 8 clusters. Table 1 gives the medoid of each cluster and the distribution of the population regarding the various clusters. Cluster validity values are given table 2.

The *cluster 1* corresponds to alcohol (≥14 IDU) and tobacco consumers, never treated and represents approximately 32% of population. The *cluster 2* corresponds to alcohol (≥14 IDU) and tobacco consumers but already treated (not the first time) and represents approximately 24% of population. The *cluster 3* corresponds to alcohol (<14 IDU) and tobacco consumers never treated and represents approximately 11% of population. The *cluster 4* corresponds to alcohol (≥14 IDU) and tobacco consumers, already treated but having also other drug dependence (other than cannabis). It represents approximately 8% of population. The *cluster 5* corresponds to alcohol (≥14 IDU), tobacco and cannabis consumers but already treated. It represents approximately 8% of population. The *cluster 6* corresponds to the poly-addict consumers (alcohol, tobacco, cannabis, substitute opiate, other drugs) never treated and represents approximately 4% of population. The *cluster 7* corresponds to alcohol (≥14 IDU) and tobacco consumers with behavioural addiction without eating disorders but already treated. It represents approximately 6% of population. The *cluster 8* corresponds to the poly-addict consumers (alcohol, tobacco, cannabis, other drugs) without substitute opiate and already treated. It represents approximately 8% of population treated in aftercare centre for addiction (in France).

Table 1. The medoid of each cluster and percentage of population (301 observations)

	Var 1	Var 2	Var 3	Var 4	Var 5	Var 6	Var 7	%
Medoid - Cluster 1	1	0	1	0	0	0	0	32%
Medoid - Cluster 2	1	1	1	0	0	0	0	24%
Medoid - Cluster 3	0	1	1	0	0	0	0	11%
Medoid - Cluster 4	1	1	1	0	0	1	0	8%
Medoid - Cluster 5	1	1	1	1	0	0	0	8%
Medoid - Cluster 6	1	0	1	1	1	2	0	4%
Medoid - Cluster 7	1	1	1	0	0	0	1	6%
Medoid - Cluster 8	1	1	1	1	0	1	0	8%

Table 2. Cluster validity values

Dunn Index	Davies-Bouldin's index	Silhouette index
1.069	0.797	0.526

4.2 Psychiatric

Concerning psychiatric factor, the following variables are observed:
Var 8: Depression,
Var 9: Anxious,
Var 10: Personality disorders,
Var 11: Psychotic disorders,
Var 12: Hospitalization in psychiatric centre,
Var 13: Suicide attempt,
Var 14: Eating disorders.

We find 9 clusters. Table 3 gives the medoid of each cluster and the distribution of the population regarding the various clusters. The cluster validity values are given table 4. From the table 3, we observe that roughly 16% of the population do not have any psychiatric problem (cluster 2). Consequently, patients suffering from addictions have serious psychiatric illness. The quality (Silhouette index= 0.166) of the cluster 1 is not very good because *Var 1* and *Var 2* are not discriminant in this cluster. The clinical validation of theses nine clusters is presented in the next section.

Table 3. The medoid of each cluster and percentage of population (301 observations)

	Var 1	Var 2	Var 3	Var 4	Var 5	Var 6	Var 7	%
Medoid - Cluster 1	1	1	1	0	1	1	0	12%
Medoid - Cluster 2	0	0	0	0	0	0	0	16%
Medoid - Cluster 3	1	1	1	0	1	1	2	6%
Medoid - Cluster 4	1	0	0	0	1	0	0	8%
Medoid - Cluster 5	0	1	0	0	0	0	0	11%
Medoid - Cluster 6	1	0	0	2	1	0	0	7%
Medoid - Cluster 7	1	0	0	0	0	0	0	14%
Medoid - Cluster 8	1	0	0	0	0	1	0	14%
Medoid - Cluster 9	1	1	2	0	0	0	0	12%

Table 4. Cluster validity values

Dunn Index	Davies-Bouldin's index	Silhouette index
0.706	0.922	0.403

The clinical validation concerns psychiatric disorders from the features of each cluster presented in the table 3. It is proposed by the resident doctor (C. Digonnet - psychiatrist) of the aftercare centre for addictions situated France.

In a first step, the psychiatrist reads the table 3 in column. On this manner, he identifies what are the variables with item value set to 2 (acuity ++) identifying the visible signs of the disease. The first cluster which draws attention is the cluster 3. This cluster vouches for the accuracy of a fact that in the observed population, *eating disorder* is coupled with *personality disorders* with *previous history* (except psychotic disorders). It represents 6% of the observed population. The second is *Var 4 "Psychotic disorders"* and corresponds to the cluster 6. It is not a surprise for the resident doctor. The cluster 6 corresponds to psychopathic people institutionalized with depressive phase. The cluster corresponds well to a clinical reality. This profile concerns 7% of the observed population. The third is *Var 3 "Personality disorders"* and corresponds to the cluster 9. This cluster corresponds to borderline personality with depression and anxiety disorders. This profile concerns 12% of the observed population. *Var 6 "Suicide attempt"* is equal to 1 in three clusters and is always in relation with *Var 1 "depression disorders"*:

- Cluster 3 corresponds to seriously ill people without psychotic and eating disorders,
- Cluster 1 corresponds to seriously ill people including eating disorders,
- Cluster 8 corresponds to dysthymic people.

These 3 clusters correspond to 32% of the observed population.

In a second step, the resident doctor analyse the result in global in order to identify the critical cases. Clusters 1 and 3 corresponds to the seriously psychiatric ill people institutionalized (18%) in which the cluster 6 is added (6% of psychotics).

5 Conclusion

The article presents a new clustering analysis method based on coupling b-coloring of graph and Bregman hard clustering algorithm. The advantage of the b-coloring is to automatically find the number of clusters and to identify the "best representative object" of each cluster. In that way, we optimize the choice of these medoids regarding to Bregman hard clustering. We try out the approach on real data concerning a patient cohort suffering from addictions. Data are been collected in seven different aftercare centres for addictions located in France. The obtained clustering is satisfying regarding cluster validity values. The obtained clusters are also analyzed by the resident doctor (psychiatrist) of the aftercare centre for addictions located in Saint-Galmier (France).

There are many interesting issues to pursue. This work is a first approach. Automatically determining the number of clusters has been one of the most difficult problems in data clustering. Usually, clustering algorithms are run with different values of *k* and the best value of *k* is chosen based on various predefined criterion. We also have to validate our approach regarding these classical approaches such as bisecting k-means, x-means, G-means, fuzzy c-means...

We limited this work to crisp clustering. But, many real world datasets and more particularly addictions have inherently overlapping between clusters meaning that one observation can potentially belong to more than one cluster. Logistic regression is mainly used in medical domain. It allows in the one hand, to identify what are factors (variables) characterizing each cluster, on the other hand, to estimate the probabilities regarding the factors of each observation in each cluster. So, it is possible to improve the hard clustering in assigning data points to multiple clusters. It will be also interesting to compare the results with soft clustering (density) or overlapping clustering approaches.

Acknowledgments. The author would like to acknowledge Mr. F. Navarro (Chairman of the Board of "Mutualité Française" Rhône-Alpes - France), as well as all the staff who had the kind enough to entrust us this project, data to validate our models and who answered our numerous questions: the resident Doctor C. Digonnet (CMA-MFL Saint-Galmier - France) and the F.N.E.S.A.A (Fédération Nationale des Établissements de Soins et d'Accompagnement en Addictologie), a French union of healthcare rehabilitation institutions for dependence/addiction. The authors are very grateful to the reviewers for their comments which were both useful and helpful.

References

1. Banerjee, A., Merugu, S., Dhillon, I.S., Ghosh, J.: Clustering with Bregman divergences. Journal of Machine Learning Research 6, 1705–1749 (2005)
2. Davies, D.L., Bouldin, D.W.: A cluster separation measure. IEEE Transactions on Pattern Analysis and Machine Intelligence 1(4), 224–227 (1979)
3. Diestel, R.: Graph Theory. Springer, Heidelberg (2006)
4. Dunn, J.C.: Well separated clusters and optimal fuzzy partitions. Journal of Cybernetics 4(1), 95–104 (1974)
5. Elghazel, H., Deslandres, V., Hacid, M.-S., Dussauchoy, A., Kheddouci, H.: A new clustering approach for symbolic data and its validation: Application to the healthcare data. In: Esposito, F., Raś, Z.W., Malerba, D., Semeraro, G. (eds.) ISMIS 2006. LNCS (LNAI), vol. 4203, pp. 473–482. Springer, Heidelberg (2006)
6. Elghazel, H., Kheddouci, H., Deslandres, V., Dussauchoy, A.: A Graph b-coloring Framework for Data Clustering. Journal of Mathematical Modelling and Algorithms 7(4), 389–423 (2008)
7. Irving, W., Manlove, D.F.: The b-chromatic number of a graph. Discrete Applied Mathematics 91, 127–141 (1999)
8. Li, T., Ma, S., Ogihara, M.: Entropy-Based Criterion in Categorical Clustering. In: Proceedings of the 21st International Conference on Machine Learning, Banff, Canada (2004)

9. Ogino, H., Yoshida, T.: Toward Improving Re-coloring Based Clustering with Graph b-Coloring. In: Zhang, B.-T., Orgun, M.A. (eds.) PRICAI 2010. LNCS, vol. 6230, pp. 206–218. Springer, Heidelberg (2010)
10. Rousseeuw, P.J.: Silhouettes: a graphical aid to the interpretation and validation of cluster analysis. Journal of Computational and Applied Mathematics 20, 53–65 (1987)
11. Walesiak, M.: The Generalized Distance Measure. In: Multidimensional Statistical Analysis. Wrocław Academy of Commerce, Wrocław (2002)

Womb Grammars as a Bio-inspired Model for Grammar Induction

Leonor Becerra-Bonache[1], Verónica Dahl[2], and M. Dolores Jiménez-López[3]

[1] Laboratoire Hubert Curien, Jean Monnet University, 18 rue Benoit Lauras, 42100, Saint-Etienne, France
leonor.becerra@univ-st-etienne.fr
[2] Simon Fraser University, Burnaby, BC V5A-1S6, Canada
veronica@cs.sfu.ca
[3] Research Group on Mathematical Linguistics, Universitat Rovira i Virgili, Av. Catalunya 35, 43002 Tarragona, Spain
mariadolores.jimenez@urv.cat

Abstract. Womb Grammars are a recently developed constraint-based framework designed for inducing the syntax of a human language from that of another, described in terms of properties [8]. They have been named by analogy to the generative power of human wombs, capable of engendering any race given appropriate input. In this article we integrate various facets of Womb Grammars from this bio-inspired viewpoint, and we extract from the various applications in which they have been used, the set of features and advantages that make them a unique tool among contemporary language processors. We also discuss extensions and limitations.

1 Introduction

We live within an *information society* in which there is a need of having a linguistic technology to deal with information. It is here where we find the so-called *language technologies*, defined as the application of natural language knowledge to the development of computer systems that are able to recognize, understand and generate natural language in various forms. The benefits of using such systems are innumerable for our society; among others, the interaction between humans and computers could be really improved. Although important advances have been achieved in this field, there is still a lot of room for enhancements.

The technological advances that our society has experienced in the last decades has also had an impact in our social lives. The development of *social media technologies*, such as social networks, internet forums, etc. has revolutionized the way in which humans communicate. We live in a society where, thanks to the Internet revolution, long distances seem shortened and new methods of socializing have been developed. For example, people can now share information in virtual communities and networks with just one click. The use of social media has increased in the last years and here language technologies are being used to provide deep analysis of the content generated by the users. Taking into account that the texts written

J.B. Perez et al. (eds.), *Trends in Practical Applications of Heterogeneous Multi-agent Systems. The PAAMS Collection*, Advances in Intelligent Systems and Computing 293,
DOI: 10.1007/978-3-319-07476-4_10,

by users often contain grammatical errors, it is really necessary for researchers to develop automated tools to process these "imperfect" texts.

Therefore, it is of great interest to develop systems that are not only able to understand natural language, but also able to process the text, even if it is not correct (from a grammatical point of view). In this context, Verónica Dahl and Emilio Miralles [8] have recently proposed a new constraint based paradigm called *Womb Grammars*. They were designed to induce, from simple phrases of a given language, the grammar of another language. They present a unique approach to grammar induction: that of weeding out constraints of a hybrid or a universal constraint-based grammar. Thus they depart from the three contemporary approaches to grammar induction, which view grammar either as a) a parametrized, generative process explaining the data [14], b) a probability model, so that learning a grammar amounts to selecting a model from a prespecified model family [5], or c) a Bayesian model of machine learning [13].

In this paper, we provide a *bio-inspired* presentation of this novel model, and we analyze the main features and advantages that make it unique among state-of-the-art language processors.

2 Womb Grammars

2.1 Motivation

In her invited talk *Agents for Linguistic Capture* [15], Veronica Dahl noted that the problem of engendering all grammars has some similarities to nature's problem of engendering all races. Nature adopted a brilliant solution to this problem: a generative apparatus -the human womb- can engender all races given an appropriate input (i.e., a fertilized egg). How can we achieve a similar economy of means for the parallel problem of engendering all grammars?

Continuing with this metaphor inspired from biology, we could look at trying to identify what *input* would be needed, what *process* would be necessary and what would be the *output*.

- **Input.** We can distinguish two types of inputs:
  1) The *egg* would be the *source language's syntax*. If we want to come up with the grammar of a language we are targeting and we have a similar language that we know the syntax of, then how can we use the language that we know to help find the grammar for the language we don't know about? We can consider that the source language's syntax is known because that is the language we know.
  2) The *sperm* would be the *target language's lexicon* and the *target language's representative corpus*. We also would know what the target language's lexicon is, together with the target language's representative corpus (we look at sentences from the target language that we know are correct. We can come up with a corpus of correct sentences in the language with the help of informants of the target language: they can tell us what lexicon is, they just don't know what the syntax is. They can use it but they cannot describe it).

- **Process.** In order to have the output we want, which is the target language's syntax, we go through a process that would be similar to *gestation*. The goal here would be a parser for the source language's syntax, which would not only be capable as usual of parsing the source language, but also capable of accepting another language's lexicon and of parsing its corpus even though foreign. So, we take the parser of the language we know, and we run it with the lexicon and the corpus of the language we want to know the grammar of. We will call this parser a *linguistic womb*.
- **Output.** The *baby* would be the *target language's syntax.*

2.2 The General Problem

The general problem would be to derive the grammar based on correct input sentences, where the source syntax and lexicon are known, the target lexicon is known, and we must infer the *target syntax.*

A possible solution would be the following. If we run a language's parser with the lexicon and corpus of another language (it has to be a very robust parser, because if not it is going to fail), then what it would produce as embryo will be a list of violated properties (for instance, if an adjective must precede the noun, and in the target language is the other way round, the precedence property of the adjective to the noun will be violated and this is going to come up as the result from the parser). Since the corpus is correct for the target language (this is a requirement we make), the properties output as violated are actually correct properties for the target language. Thus, the parser must modify its own grammar to accommodate these properties, resulting in the baby: a correct grammar for the target language. The question is: how to implement this solution?

2.3 Linguistic Wombs

Constraint Based Grammars that use properties define the acceptability of a sentence in terms of constraint satisfaction between pairs of constituents. For example, within a noun phrase the following such constraints hold: constituency (e.g. determiners, adjectives and nouns are allowable constituents), linear precedence (determiner precedes adjective, determiner precedes noun, adjective precedes noun), unicity (only one determiner is allowed), etc. Therefore, in this constraint-based paradigm, a sentence is characterized through the list of constraints a phrase satisfies and the list of constraints it violates. It follows that not only correct sentences can be handled.

Womb Grammars are a novel constraint-based framework introduced by Veronica Dahl and Emilio Miralles in [8]. In contrast to traditional constraint-based linguistic models such as Head-driven Phrase Structure Grammars [18] or Property Grammars [4], Womb Grammars are not only useful for characterizing the acceptability of a sentence in terms of linguistic constraints, but also for *inducing* a target language's constraints from those of a source language. The reason is that they were conceived for grammar induction rather than only for parsing sentences. Two different approaches have been developed:

– **Hybrid Approach.** This first approach uses as the source language a sufficiently studied one, and the target language is a less studied but related one. How does it work exactly?
  - The input is the target corpus, the target lexicon and the source syntax.
  - This input is given to a Womb Grammar parser. Because it is parsing correct sentences but only with respect to a target, some of the properties that the source syntax describes are going to be violated.
  - The parser returns a list of violated syntax properties, that it founds by parsing the entire corpus, sentence by sentence.
  - This list of violated syntax properties goes as input to a module or an agent that repairs the grammar of the source, to make the corpus of sentences accepted. So it corrects the grammar and then produces the target syntax.

  For example, imagine that English is the source language and Spanish is the target, and we assume that English adjectives precede their noun. If we then enter in the corpus a sentence like "el coche verde" [1], this precedence property will be violated. Taking into account the list of violated properties, the grammar repairing module can modify the constraints in the source syntax so that none are violated by the target corpus, and produce the missing syntactic components of the target syntax's parser.

– **Universal Approach.** This second approach considers a universal source of language. Hence, instead of a related language, a universal grammar is used that simply consists of all possible grammar properties for the constituents given, even contradictory ones (for instance, determiner precedes noun, noun precedes determiner). Then, the constraints left over from a run through a representative corpus of any language will contain only properties that are not violated by the target language. The process is simplified as follows:
  - The input is the target corpus, the target lexicon and the universal syntax (source language is now the universal language).
  - We run the target corpus and the target lexicon with the universal syntax.
  - We eliminate all grammar properties that are violated.
  - The remaining constraints describe the grammar.

In terms of our bio-inspired metaphor, the move into universal grammar in the second approach is analogous to fertilization in vitro. While in normal fertilization the input (egg and sperm) is fixed in the sense that the egg is produced by the same organism which processes both of them, with fertilization in vitro any womb becomes a super womb capable of receiving an already fertilized egg, in other words, of truly generating all races, since for each race, an appropriate pair of egg and sperm could be found. The way the universal grammar approach

[1] Literally, "the car green".

materializes this metaphor is akin to having all possible types of eggs and sperm available, and discarding those clashing with desired characteristics having been specified a priori.

It is worth noting that Womb Grammars have also been successfully applied in various domains, such as grammar induction for under-studied human languages [8], language acquisition [9] and second language learning [3]. To a certain extent, the Womb Grammar framework can perhaps be transposed to new applications such as formal languages, since in many cases these include similar types of constraints, e.g. the formal language of n a's followed by n b's exhibits a precedence constraint (a must precede b). Further research could investigate what kinds of constraints are proper of formal languages, and how to extend Womb Grammars with these, so that we can adopt the same scheme for inferring a target formal grammar from another one. It could be trickier however to accept incomplete or erroneous input in formal languages, since errors could be more easily construed as legitimate parts of the new grammar.

3 Advantages of Womb Grammars with Respect to Other Formalisms

Womb Grammars present several advantages with respect to other formalisms in many respects. In this section we try to highlight some of those advantageous features.

From the point of view of their application to Natural Language Processing, advantageous features of Womb Grammars are: the fact of being language-independent and their capacity to parse non-canonical input.

The Womb Grammar formalism is not language-specific, but the same mechanism can be applied by and large to any natural language. This characteristic offers the possibility of having just one parser for any language, solving the problem of adaptation to different languages of parsers developed for English: while it is possible that for some families of language we may need to consider specific properties not present in English, the fact that properties themselves are modular makes it possible to do so without affecting the methodology itself. The typological diversity of natural languages poses important challenges for the techniques used in machine translation, syntactic parsing and other areas of natural language processing. It is well known that English is the main language of focus for the computational linguistics community. Unfortunately, while progress on parsing English has inspired some advances on other languages, it has not yielded high-quality parsing. Models developed for English do not necessarily perform well for languages typologically different from English. As stated by Nivre [16], several studies have reported a substantial increase in error rate when applying state-of-the-art statistical parsers developed for English to other languages, such as Czech, Chinese, German or Italian. Those problems, led to the idea of developing language-independent systems for parsing. The MaltParser developed by Nivre et al. [16] is an example of this type of system. It can be defined as a data-driven dependency parsing that can be used to induce a parser for a new language from a treebank sample in a simple flexible manner. Womb Grammars are also a system that can adapt to any language, since they can use a representative sample input set to generate the corresponding grammar.

Womb Grammars are able to parse incorrect or incomplete input, this is, the so-called noisy text. On the contrary, most of the traditional parsing approaches fail when the input is not correct. Recently, one of the most interesting lines of research to be addressed in the area of NLP is the understanding of social media content [12,10]. Traditional Natural Language Processing methods fail when try to process the typology of text in the social media, due its inherently noisy nature. In social media, we find a free-form nature of language, encompassing spelling inconsistencies, a free-form adoption of new terms and jargon, and violations of grammar norms [2]. Those characteristics make parsing of data from the web a notoriously difficult task, since parsers are generally trained on more canonical data. Taking into account that most applications that rely on parsing –such as machine translation and information extraction– need to handle very often unedited text coming from domains on the web such as blogs, discussion forums, social networks, etc., it is important to make progress in parsing such types of texts [17]. Therefore, parsers must improve performance on text which is mildly ungrammatical; this is text which contains grammatical errors that are produced by speakers of a language. In order to reliably translate and extract information from the web, it is necessary to develop formalisms able to extract the syntactic structure of non-canonical language. As stated by Forster et al. [12], the goal is not to detect and correct the error, but rather to ignore it and, therefore, be able to parse it. Womb Grammars are able to parse both grammatical and ungrammatical sentences, so they can be seen as a useful tool for many language technology applications, including technologies applied to social media.

From the point of view of their theoretical value in linguistics, we highlight the capacity that Womb Grammars have to distinguish different levels of grammaticality. Womb Grammars characterize a sentence through a list of satisfied constraints and a list of violated constraints. By evaluating those results, one can calculate the level of grammaticality of a sentence. If we look at the history of linguistics, the concept of levels of grammaticality/acceptability has been always present [1]. However, although the different linguistic schools have considered the possibility to deal with the idea of graduality within grammatical issues, it is difficult to find a linguistic theory that actually considers grammar as a non-discrete object. It is therefore difficult to find theoretical models in linguistics that provide formal tools (computationally adequate) to give a clear account of the conditions that determine the degree of grammaticality of a given structure. During the last years, there is a great interest to introduce the idea of gradience in grammar [11]. Following this research line, we could propose Womb Grammars as a formal tool for defining the idea of fuzzy grammaticality in linguistics. A simple measure of fuzziness could be given by the number of violated constraints; a more sophisticated one could include preference measures for the different constraints (e.g. the requirement for a phrase's head would be more important a constraint than the requirement of its agreeing in gender and number with its modifiers).

Finally, from the point of view of its motivation, Womb Grammars present the advantageous feature of being inspired by the process of children's language acquisition. Following Chomsky's ideas, we are born with the innate capacity to learn a language, and by being exposed to any language, we are able to speak. According to Government and Binding theory proposed by Chomsky [7], in a natural language we have to distinguish between principles and parameters: principles are built-in to the human mind, while parameters are set from the data the child encounters. So, the only difference between people speaking Spanish and people speaking French is that they have been exposed to different language evidence that has set the parameters of the Universal Grammar [6] differently. We can consider that a Womb Grammar consists of a Universal Grammar composed by principles that are the same for any natural language. The Womb Grammar processor will set the parameters depending on the data/input it parses, thus generating different grammars for different languages, but starting always from the same formalism.

Summing up, the motivation and the features of Womb Grammars make of this formalism an interesting framework to be considered not only in the area of Natural Language Processing, where interesting results can be obtained from its application of parsing noisy text, but also in the theoretical analysis of language, by providing to linguists elegant and formal tools to account for and comparatively test central problems in Linguistics.

4 Conclusion

We have discussed, under the light of its bio-inspired features, the unique approach of Womb Grammars to syntax induction. We have also discussed their main advantages, and pointed out some possible further extensions and new applications. We should be careful to stress that while the results obtained for simple phrases have been good within the applications of Womb Grammars done to date, an extension of this paradigm from proof-of-concept into a fairly complete, real world application with wide language coverage remains to be done. With this paper we hope to have shed light on the bio-inspired aspects of this intriguing paradigm, and to stimulate further research along these lines.

Acknowledgment. The work of Verónica Dahl was supported by NSERC grant 31611024.

References

1. Aarts, B.: Conceptions of gradience in the history of linguistics. Language Sciences 26, 343–389 (2004a)
2. Baldwin, T.: Social Media: Friend or Foe of Natural Language Processing? In: The 26th Pacific Asia Conference on Language, Information and Computation, pp. 58–59 (2012)

3. Becerra-Bonache, L., Dahl, V., Miralles, E.: On Second Language Tutoring through Womb Grammars. In: The 12th International Work-Conference on Artificial Neural Networks (IWAAN), pp. 189–197 (2013)
4. Blache, P.: Property grammars: A fully constraint-based theory. In: Christiansen, H., Skadhauge, P.R., Villadsen, J. (eds.) CSLP 2005. LNCS (LNAI), vol. 3438, pp. 1–16. Springer, Heidelberg (2005)
5. Cohen, S.B., Smith, N.A.: Covariance in unsupervised learning of probabilistic grammars. Journal of Machine Learning Research 11, 3017–3051 (2010)
6. Cook, V.J.: Chomsky's Universal Grammar: An Introduction. Blackwell (1988)
7. Chomsky, N.: Language and Problems of Knowledge: The Managua Lectures. MIT Press (1988)
8. Dahl, V., Miralles, E.: Womb grammars: Constraint solving for grammar induction. In: The 9th International Workshop on Constraint Handling Rules, pp. 32–40 (2012)
9. Dahl, V., Miralles, E., Becerra-Bonache, L.: On Language Acquisition Through Womb Grammars. In: The 7th International Workshop on Constraint Solving and Language Processing, pp. 99–105 (2012)
10. Derczynski, L., Ritter, A., Clark, S., Bontcheva, K.: Twitter Part-of-Speech Tagging for All: Overcoming Sparse and Noisy Data. In: Proceedings of Recent Advances in Natural Language Processing, Hissar, pp. 198–206 (2013)
11. Fanselow, G., Fery, C., Vogel, R., Schelsewsky, M. (eds.): Gradience in Grammar: Generative Perspectives. Oxford University Press (2006)
12. Foster, J., Wagner, J., van Genabith, J.: Adapting a WSJ-Trained Parser to Grammatically Noisy Text. In: Proceedings of the 46th Annual Meeting of the Association for Computational Linguistics on Human Language Technologies: Short Papers, pp. 221–224 (2008)
13. Headden, W.P., Johnson, M., McClosky, D.: Improving unsupervised dependency parsing with richer contexts and smoothing. In: HLT-NAACL, pp. 101–109 (2009)
14. Klein, D., Manning, C.D.: Corpus-based induction of syntactic structure: Models of dependency and constituency. In: Scott, D., Daelemans, W., Walker, M.A. (eds.) ACL, pp. 478–485 (2004)
15. LAFLang: Learning, Agents and Formal Languages, Special Session. In: ICAART 2013 (International Conference on Agents and Artificial Intelligence), Barcelona, Spain (2013)
16. Nivre, J., Jensnilsson, J., Chanev, A., Eryigit, G., Kubler, S., Marinov, S., Marsi, E.: MaltParser: A language-independent system for data-driven dependency parsing. Natural Language Engineering 13(2), 95–135 (2007)
17. Petrov, S., McDonald, R.: Overview of the 2012 Shared Task on Parsing the Web. In: Workshop on the Syntactic Analysis of Non-Canonical Language (SANCL 2012), Montreal (2012)
18. Pollard, C., Sag, I.A.: Head-driven Phrase Structure Grammars. CSLI, Chicago University Press, Chicago (1994)

A Multiagent Approach Using Model-Based Predictive Control for an Irrigation Canal

Van Thang Pham, Clément Raïevsky, and Jean-Paul Jamont

Université de Grenoble Alpes, Laboratoire LCIS, Valence, France
firstname.lastname@lcis.grenoble-inp.fr

Abstract. This paper presents the application of the multiagent paradigm to a distributed model-based predictive control (DMPC) scheme in order to improve its fault tolerance, give it the ability to dynamically adapt its strategy to optimize energy consumption, and to allow it to scale up. This approach is illustrated in the control of a canal simulated using realistic, physics-based 1D models in MatLab. The individual agent behavior, based on DMPC, and the multiagent composition mechanism are described. Presented simulation results illustrate the ability of the proposed control scheme to adapt to a hardware failure and to take global strategies into account.

Keywords: distributed control, water system, predictive control.

1 Introduction

Water systems are complex collections of interconnected water bodies (e.g. lakes, reservoirs), natural channels, pipes, and canals. Gates, dams, pumps, gates, and valves are used to control water flows and achieve proper delivery of irrigation water. Controlling such heterogeneous, distributed, and dynamic systems is extremely challenging using centralized techniques because of the constraints these techniques pose on communication, computation cost, and adaptivity.

We focus our study on man-made canals which convey water to remote users. The main problem is to ensure that the appropriate amount of water is available to the user at the right time. The associated controlling system is thus responsible for adjusting water level in canals to satisfy water demand while minimizing wastes. In this paper we therefore consider distributed control scheme and especially how the multiagent paradigm can be combined with model-based predictive control to achieve an efficient and adaptive management of a water system.

The next section describes the problem and existing work related to this use case, in automation and multiagent communities. Section 3 presents the way we addressed the problem and especially the agents' individual behaviors and the agent composition mechanism. Simulation protocol, results and discussion are exposed in Section 4.

J.B. Perez et al. (eds.), *Trends in Practical Applications of Heterogeneous Multi-agent Systems. The PAAMS Collection,* Advances in Intelligent Systems and Computing 293,
DOI: 10.1007/978-3-319-07476-4_11, © Springer International Publishing Switzerland 2014

2 Problem Description

The use case We consider in this study an irrigation channel consisting of N cascaded pools. Each pool is usually described by a set of two partial differential equations (PDEs) named Saint-Venant equations, which represent the mass and the momentum conservation :

$$\begin{aligned} &B_i \partial_t h_i + \partial_x Q_i && = 0 \\ &\partial_t Q_i + \partial_x \left(\frac{Q_i^2}{B_i h_i} + \frac{1}{2} B_i g h_i^2 \right) && = g B_i h_i (I_i - J(Q_i, h_i)), \\ &(x,t) \in [0, L] \times [0, \infty), && i = 1, ..., n, \end{aligned} \tag{1}$$

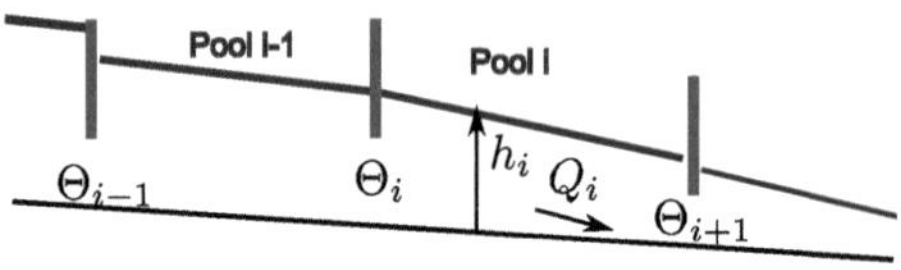

Fig. 1. Multi-pool system

where h_i denotes the water depth, Q_i the discharge, g the gravitational acceleration, B_i the channel width, I_i the slope, and J the friction term. Here t and x classically stand for time and space coordinates, and ∂_t, ∂_x denote the partial derivatives w.r.t. t, x respectively.
Interconnections between pools are subject to a set of $N+1$ sliding gate equations:

$$Q_{g_i} = K_i^2 \Theta_i^2(t) 2g(h_{us}^{g_i} - h_{ds}^{g_i}), i = 1, ..., N+1, \tag{2}$$

and $N-1$ discharge conservation constraints:

$$Q_i(L,t) = Q_{i+1}(0,t), \quad i = 1, ..., n-1, \tag{3}$$

where Q_{g_i} is the discharge through gate i, K_i is the gate coefficient, Θ_i its opening, $h_{us}^{g_i}$ and $h_{ds}^{g_i}$ are the water levels in upstream and downstream pools respectively.

Related Work. The control problem of this complex system is a challenging task. Using a centralized approach to control such systems as a whole is very difficult because of the high computational requirements. This obstacle can be tackled by using the so-called distributed strategy in which the entire system is divided into several sub-problems, each of them being allocated to a local agent. Each agent can choose among a large number of control strategies to determine its behavior regarding the control of its actuators [1,2]. In this study, we chose to use the Model-based Predictive Control (MPC) because of its ability to handle constraints, to incorporate forecast information and to minimize a given criterion [3,4,5]. This approach can now be found in a wide variety of application areas including chemicals, food processing, automotive, aerospace and metallurgy [6]. Based on measurements obtained at time t, the controller uses a model of the system to predict its future behavior over a prediction horizon T and determines the action such that a predetermined cost function is optimized. Due to disturbances and model-plant mismatch, the true system behavior is different from the predicted behavior. In order to incorporate some feedback mechanism, the obtained optimal solution will be implemented only until the next measurement becomes available, e.g. at time $t+\sigma$. Using this new information, the whole procedure "prediction and optimization" is repeated to find a new action with the prediction horizon moving forward, which explains why this approach is also termed Receding Horizon Control.

Several pieces of work propose multiagent architectures in the context of control of irrigation systems. These works cover issues ranging from river system data collection [7,8] to data post processing (like flooding anticipation [9]) and simulation of critical scenarios to support decision-making [10]. In [11], authors propose a multiagent architecture to diagnose faults along water canals. A multiagent approach has been proposed to model the evolution of the mutual influences between consumption and availability of water resources in a semi-arid river basin [12].

Conclusion. Automation offers rigorous approaches guided by the physical models of (natural) system to monitor. Distributed approaches enable to address high computational requirements. Multiagent approaches have been proposed to address various problems in the context of river monitoring.
Our aim is to use a multiagent approach in order to benefit from the advantages of decentralization while keeping the benefits from automatic control. In other words, our goal is to design a control system which: (1) is more responsive to unexpected events (i.e. has a satisfying fault tolerance), (2) takes into account complex strategies related to energy management and alert management, and (3) addresses the scaling problem.

3 Our Approach

ASTRO Architecture. Each gate is controlled by an agent which uses the ASTRO architecture [13]. ASTRO is especially adapted to a real time context. The integration of deliberative and reactive capabilities is possible through the use of parallelism in the structure of the agent. Separating Reasoning/Adaptation and Perception/Communication tasks allows a continuous supervision of the evolution of the environment. The reasoning model of this agent is based on the Perception/Decision/Reasoning/Action paradigm. The cognitive reasoning is thus preserved, and predicted events contribute to the normal progress of the decision making process. The agent's individual and collective behaviors are integrated in this architecture.

Individual Behavior. At each time step t agent i carries out the following tasks:

1. collect information regarding the state of pool i (e.g. taking water depth and discharge readings);
2. predict the evolution of the pool's state over the time horizon t to $t+T$ (by using the Saint-Venant equations (1)) and determine the control action that minimizes the distance from the perceived water level and discharge in the pool to a given profile. More precisely, the agent has to solve the following optimization problem:

$$\min_{\bar{\Theta}_i} J = \int_t^{t+T} \int_0^{L_i} A(x)(\bar{h}_i(x,\tau) - h_{0i}(x))^2 dx d\tau + B(x)(\bar{Q}_i(x,\tau) - Q_{0i}(x))^2 dx d\tau \tag{4}$$

   where the notation $\bar{\cdot}$ represents the prediction variables (internal variables used by the agent), h_{0i} and Q_{0i} are the desired level and discharge respectively, $A(x)$ and $B(x)$ are some weighting positive functions.
3. apply the obtained optimal solution until new information can be collected.

Note that the interactions between pools exist via the gate equations (2) and (3). The individual behavior described above will conduct to a local optimum if these interactions were not taken into account. To avoid these local optima, a collective behavior must be added to the agent's decision making.

Collective Behavior. Two collective behaviors are presented here: the fault adaptation and the collective solution optimization. Other behaviors such as wound alerts are not discussed for lack of space.
Adaptation In order to give adaptability to the controlling system as a whole, we added composition abilities to the agents. This ability allows agents that detect they are no longer able to control their gate to be integrated in still functioning agents.

The composition process unfolds as follows: first, the failing agent, wanting to initiate a composition, sends a composition request to the most suitable agent. This choice is based on the estimation of which composition will be able to fulfil its (now defunct) function. In our example, the most suitable agent is the upstream one. Second, upon composition acceptance from the upstream agent, the failing agent sends a topology reconfiguration message to its previously downstream agent. Third, the upstream and failing agents merge and the system resumes its controlling scheme, this new composed agent being responsible for the control of two pools.

Using this mechanism, the system can maintain its overall ability to control the water flow in the canal event though one agent lost its ability to act on the valve it controlled. Compared to a centralized control scheme or a static distributed MPC (DMPC), which would require to be modified if an actuator failure occurs, the presented control scheme is more adaptive.

Solution optimization. Agents will repeatedly exchange information to make the system converge toward the global optimum (see Fig. 2). To this end, agents must synchronize their computation as follows:

- each agent i solves the optimization problem 4 by incorporating received information from neighbors;
- after all agents have finished solving their problem, agent i sends its prediction values on water level and discharge at upstream, namely $(\bar{h}_i(0,\cdot), \bar{Q}_i(0,\cdot))$ and at downstream, namely $(\bar{h}_i(L,\cdot), \bar{Q}_i(L,\cdot))$, to the upstream $(i-1)$ and downstream $(i+1)$ neighbors respectively;
- each agent moves to the next iteration until the maximum number of iteration is reached.

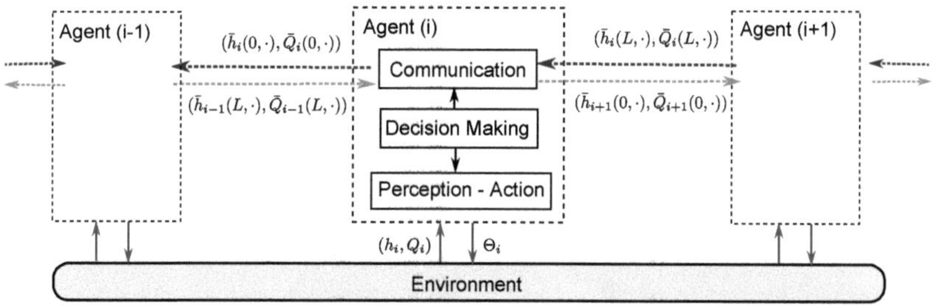

Fig. 2. Informations exchanged between agents

Notice that by using the results of [14] and [15], we can show that when the number of iterations is increased, the obtained control law is the global optimal one. The convergence of the whole system to the reference profile is then guaranteed (see [16]).

4 Evaluation

In this section, we present some simulation results in order to illustrate the effectiveness of the proposed approach.

Simulation Protocol. We consider an irrigation channel consisting of three pools and four sliding gates which are controlled by 4 agents. This system was implemented in Matlab/Simulink (see figure 4) using Saint-Venant equations (1). Agents and their communications are simulated in MASH (Mutiagent Software Hardware Simulator) which is dedicated to the design and implementation of real world embedded multi-agent systems .

Results. We performed the first simulation without any actuator failure. The results are presented in figure 3. We can observe that our multiagent MPC approach can make the system converge to the reference profile. In the second

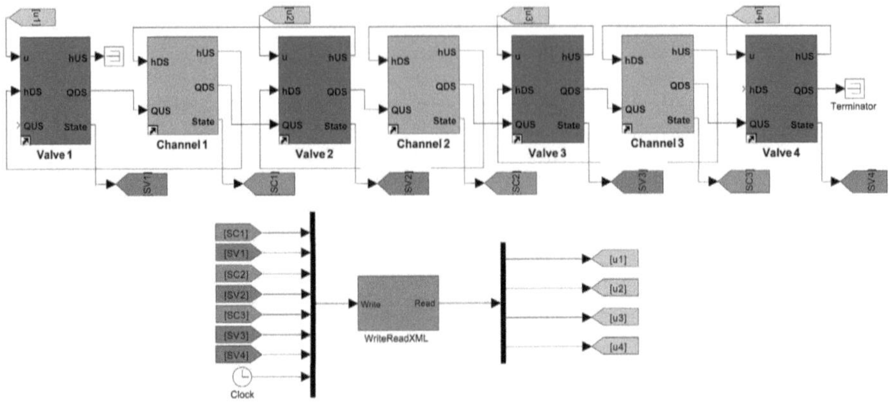

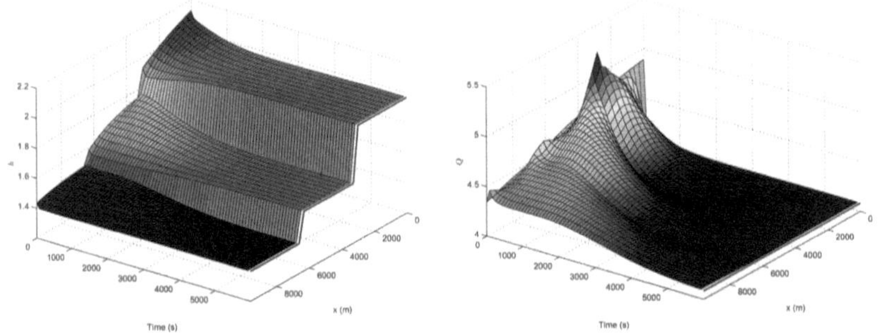

Fig. 3. Water level (left) and discharge (right) in the channel. The red lines represent the desired values

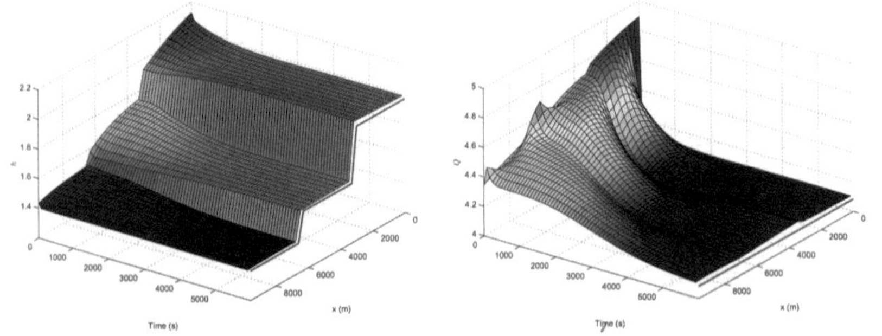

Fig. 4. Water level (left) and discharge (right) in the channel when the 2nd gate is blocked

simulation we consider the case where the second gate is blocked. The result is presented in figure 4.
We can see that water level and discharge are always regulated around the desired values despite of some small error. More complex scenarios such as an important variation of upstream water level or disturbances coming from pumping stations will be considered in a longer version of the paper.

Note also that the computation time for obtaining the decision is around $0.25s$ on an Intel Core i7 3.4GHz, 8G RAM laptop which implies that a real-time implementation is possible.
The results confirm that the proposed approach fulfill our objectives, namely:

- Increasing tolerance to fault: the composition algorithm ensures that the system resulting from a composition after a failure is still within the framework studied in [16], the convergence to a global optimum is thus guaranteed. Notice that with a centralized approach, we lost the theoretical proof of convergence.
- Possibility to incorporate complex strategy: we set the cost function of the optimization problem as being the distance between the actual profile and the desired one. Nevertheless, by taking advantages of MPC, more complex cost functions can also be used, taking into account other criteria such as energy consumption, actuator constraints, or weather forecast.
- Scaling problem: We can see that the agents' ability to make composition can be used in a recursive manner, meaning that it is possible to treat actuator faults in systems with a great number of agents. In addition, agents can also voluntarily propose compositions with neighbors sharing some common characteristics (e.g. geological regions or similar objective function). This dynamic adaptation of the MAS structure is a promising method for the decomposition problem of large-scale system which is still an open issue in the DMPC community [17].

5 Conclusion

We presented an application of the multiagent paradigm to the distributed, model-based predictive control of a water canal. In our experiments, each agent was able to control one of the canal's gates using MPC and inter agent communication in order to obtain an optimal distributed control. We simulated the failure of a gate's actuator and gave the system the ability to adapt to this failure by merging the appropriate agents into a new one. Combining the fine-grained, precise, control provided by automation methods on one hand, and strategic adaptation provided by multiagent systems on the other hand, gave our approach an optimal control scheme in favorable cases and the ability to adapt to hardware failures by dynamically modifying the multiagent system's structure. The proposed agent composition mechanism maintains the global convergence verifiability. The presented system showed promising results regarding failure tolerance, ability to take global constraints into account and aptitude to scale up.

References

1. Malaterre, P.O., Rogers, D.C., Schuurmans, J.: Classification of canal control algorithms. Journal of Irrigation and Drainage Engineering (1998)
2. Litrico, X., Fromion, V.: Modeling and Control of Hydrosystems - A Frequency Domain Approach. Springer-Verlag (2009)
3. Mayne, D.Q., Rawlings, J.B., Rao, C.V., Scokaert, P.O.M.: Constrained model predictive control: Stability and optimality. Automatica, 789–814 (2000)
4. Findeisen, R., Imsland, L., Allgöwer, F., Foss, B.A.: State and output feedback nonlinear model predictive control: An overview. Euro. J. of Control 9(3), 179–195 (2003)
5. Mayne, D.Q., Michalska, H.: Receding horizon control of nonlinear systems. IEEE Transactions on Automatic Control 35(7), 814–824 (1990)
6. Qin, S., Badgwell, T.: An overview of industrial model predictive control technology. In: Kantor, J., Garcia, C., Carnahan, B. (eds.) Fifth International Conference on Chemical Process Control - CPCV, pp. 232–256. American Institute of Chemical Engineers (1996)
7. Jamont, J.P., Occello, M., Lagréze, A.: A multiagent approach to manage communication in wireless instrumentation systems. Measurement 43(4), 489–503 (2010)
8. Matei, A.M.: Multi-agent system for monitoring and analysis prahova hydrographical basin. Technical report (2011)
9. Luo, J., Xu, L., Jamont, J.P., Zeng, L., Shi, Z.: Flood decision support system on agent grid: method and implementation. Enterprise Information Sys. 1(1), 49–68 (2007)
10. Rendón-Sallard, T., Sànchez-Marrè, M., Aulinas, M., Comas, J.: Designing a multi-agent system to simulate scenarios for decision-making in river basin systems. In: Proc. of the 9th Int. Conf. of the AI R&D, pp. 291–298. IOS Press (2006)
11. Nabais, J.L., Mendonça, L.F., Botto, M.A.: A multi-agent architecture for diagnosing simultaneous faults along water canals. In: Control Engineering Practice (2013)
12. van Oel, P.R., Krol, M.S., Hoekstra, A.Y., Taddei, R.R.: Feedback mechanisms between water availability and water use in a semi-arid river basin: A spatially explicit mas simulation approach. Environmental Modelling & Software 25(4), 433–443 (2010)
13. Occello, M., et al.: Designing organized agents for cooperation with real time constaints. In: Drogoul, A., Fukuda, T., Tambe, M. (eds.) CRW 1998. LNCS, vol. 1456, pp. 25–37. Springer, Heidelberg (1998)
14. Cohen, G., Zhu, D.L.: Decomposition coordination methods in large scale optimization problems: The nondifferentiable case and the use of augmented lagrangians. Advances in Large Scale Systems 1, 203–266 (1984)
15. Cohen, G.: Auxiliary problem principle and decomposition of optimization problems. Journal of Optimization Theory and Application 32(3) (1980)
16. Pham, V.T., Georges, D., Besançon, G.: Infinite-dimensional predictive control for hyperbolic systems. SIAM J. of Ctrl. and Optimisation (2012)
17. Scattolini, R.: Architectures for distributed and hierachical model predictive control. Journal for Process Control, 723–731 (2009)

Human-Aware Planning: A Survey Related to Joint Human-Agent Activities

Sebastian Ahrndt*, Johannes Fähndrich, and Sahin Albayrak

DAI-Laboratory, Technische Universität Berlin,
Faculty of Electrical Engineering and Computer Science,
Ernst-Reuter-Platz 7, 10587 Berlin, Germany
sebastian.ahrndt@dai-labor.de

Abstract. To become a part of a joint human-agent team, artificial agent are required to achieve joint goals with humans not only performing task for humans. This includes the ability to coordinate actions between team-members, which is e.g. addressed by Human-Aware Planning approaches. This work surveys available solutions regarding the special requirements identified for joint human-agent activities. In particular, the work concentrates on the requirement of interpredictability, which requires to include the course of actions of other team-members into the planning process of one's own course of action.

1 Introduction

Focus Human-sided self-explanation, aims to integrate users into existing systems not only as supervisors but also as a regular component and to offer the opportunity to interact with human beings, by means of the available communication channels [11]. Although the addressed systems are goal-driven and act (as much as possible) autonomously, humans have to be able to define goals, to restrict the system by means of constraints and to inspect the results of the self-organisation process [15,21]. Furthermore, humans must not only be seen as supervisors but also as agents that can be asked for help to reach specific goals [21]. These mechanisms rises significant engineering challenges for both, the decision making process of agents and the interaction between agents and human users. Concentrating on the decision making process, this work surveys requirements and available state-of-the-art solutions. In particular, we focus on the challenge to decide at which moment an interaction with humans is useful and at which situation and in which way the interventions of human users should be included into the course of actions of agents.

Motivation Autonomous agents have to decide what actions to perform, when they should be performed and—in some cases—determine which entity should be asked for help to fulfil a specific task [28]. Once the available entities do not only comprise artificial agents but also natural agents or in other words human beings, joint human-agent activities emerge. Decision making procedures

* Corresponding author.

J.B. Perez et al. (eds.), *Trends in Practical Applications of Heterogeneous Multi-agent Systems. The PAAMS Collection*, Advances in Intelligent Systems and Computing 293,
DOI: 10.1007/978-3-319-07476-4_12,

for joint human-agent activities are done by Human-Aware Planning [8] (HAP) components, that encourage the basic idea of human-agent activities, which is that agents and humans have different capabilities, advantages and disadvantages that can complement each other. One example of a human advantage is the selection of relevant information while one example of the strengths of computers is the management of vast amounts of data. Furthermore, humans excel at finding solutions to new or evolving problems, building knowledge and learning, but are less efficient and not suited for the execution of multiple operations at the same time.

To build effective team-players agents planning for such activities have to tackle the dynamic nature of humans as the human behaviour features several aspects of uncertainty for the planning process [1,16,19]. As an example, consider the fact that humans select merely good and feasible actions to fulfil a task rather then selecting the optimal ones [25]. Furthermore, after committing to a specific task humans may change their goals from one moment to another without a (for computers) comprehensible reason. These aspects of uncertainty imply a dynamical behaviour that can be seen as some kind of 'Quality of Behaviour' a human is able to provide and influences the planning process in different ways. For example, the non-optimal execution of an action influences the execution time whereas the sudden interruption of a task endangers the whole plan and might imply a replanning to reach a given goal.

One assumption of currently available human-aware planning components is that whenever a task is assigned to a human that possesses the ability to fulfil it, the human will provide results in a timely fashion. This assumption is questionable since the 'Quality of Behaviour' a human is able to provide depends on the current context of the human. For example, a human might not provide required information if he/she is currently occupied with performing another task. The primary objective of our work is to relax this assumption to a more general one. That is, whenever a task is assigned to a human with the ability to fulfil it, the human may accept or decline such task and provide results either in time or delayed [1]. This form of 'context-dependent' behaviour is essential for planning processes that account for the abilities of human beings.

Structure The remainder of this work is structured as follows. In Section 2 we give some background information about joint human-agent activities and human-aware planning components. Section 3 introduces requirements for HAP. Section 4 presents state-of-the-art HAP projects and approaches. Afterwards, in Section 5 we discuss the results and implications of this survey. Finally, Section 6 concludes our study and gives an outlook on future work.

2 Terminology

Joint Human-Agent Activities Following the definition for joint-activities given by *H.H. Clark* [10, p. 3], joint human-agent activities can be defined as an extended set of actions that is executed by an ensemble of natural and artificial agents who are coordinating with each other [10,19]. Agents cooperate with each other to overcome some certain kind of inherent limitations. Nevertheless, cooperation would be avoided if no other stimuli exist, as it adds additional cost to an activity, e.g. in terms of a communication overhead. This can be an external stimulus like a goal that is not reachable without cooperation either caused by resource or capability constraints or an internal stimulus like an extroverted character that is forced to cooperate with others [24]. These limitations can be found either on the perception level, the cognition level and/or the execution level [23,26]. As examples, consider agents with sensory malfunction (perception level), humans with a disease like dementia (cognition level) or robots that are not able to overcome obstacles like stairs (execution level). In consequence they form a symbiotic relationship in which agents perform tasks for humans and humans in return help agents [23].

Human-Aware Planning Planning procedures that account for joint human-agent activities are computed by human-aware planning components. HAP is an evolving branch of AI planning systems for collaborative settings where agents coexists with humans [8] (e.g., socially assistive robots in household environments). Following *Cirillo et al.* [9, p. 17], human-aware planning can be applied in situations in which there are artificial agents whose actions can be planned and natural agents whose action can be only predicted, sharing the same environment. Here, to build effective team-players the agents are required to include the state of the human into their planning process to anticipate the actions of the human [14]. This information then can be used to generate plans including the human as actor and respecting a set of e.g., social or interaction constraints [8,20].

3 Requirements

Advances in AI stimulate the development of more complex teamwork scenarios then those of these days, in which artificial agents become part of the team itself [22]. *Klein et al.* [18,19] emphasises that such team-work requires the participants at least to enter into an agreement to work together (named Basic Compact), to be mutually predictable and directable and to maintain a common ground. Furthermore, the authors formulate several other challenges like observability and cost control, most of them beyond the scope of this work. However, two of the challenges—namely collaboration and interpredictability—are of particular interest. Collaboration addresses the decision making process of agents—the actual planning. It includes the ability to understand and solve a problem in an incremental fashion and negotiate the course of actions with other team-members. Requirements for such planning components include the abilities to [5,6,19]:

- reveal the current status of the overall plan,
- detect possible failures during plan execution and to replan if failure occur,

- evaluate the viability of plan changes,
- replan in situations where an individual agent's capabilities are outperformed,
- recruit more capable agents to perform the replanning,
- manage retasking when plan changes occur, and
- adjust the communication capability to the agents' capabilities (human agents require UIs where software agents expect procedural calls).

Additionally, we can identify some real time constraints. If the agent application takes too much time to act on behalf of a user (e.g., for plan generation), the human might feel misunderstood or might think that the application has malfunction [20]. These stop-and-go like interaction avoids the fluent meshing of actions, which is typically for good teamwork [13].

Furthermore, we have to consider the context-dependent behaviour of humans. That means, that whenever a human is predicted to fulfil a task, the human may perform this task or not and provides results either in time or delayed [1]. This is also called interpredictability and requires to plan actions considering the actions of others [6,18]. For this, human-behaviour models either hand-crafted or derived from psychological studies can be used [16,17]. Such models define the possible behaviour, capabilities and habits of humans and should be updated during runtime to adapt to the individuals that interact with the system.

4 Related Work

Several of the presented requirements can be imported 'out of the box' from state-of-the-art approaches either from the field of human-aware planning or from adjacent research fields [27]. For example, the ability to reveal the current status of the overall plan and to detect possible failures during the plan execution is given by the 3-layer architecture of current dynamic planning components [12, p. 9]. Here, a closed loop between the planning level, the monitoring level and the execution/controller level enables interleaved planning and execution, which is utilised for plan supervision, plan revision and replanning.

This multi-layered structure is for example used in the HAP framework presented by *Cirillo et al.* [8]. The approach describes the use of intention models to decide whether an agent is allowed to perform its task or if the agent would disturb the human user and should not perform the task now. This information is used to postpone agent tasks to a more acceptable time frame. This use case is interesting but beyond the scope of our work as it is our intention to consider the abilities of humans when planning and to assign tasks to them.

Another HAP approach—the Human Aware Task Planner (HATP)—is presented by *Montreuil et al.* [20]. HATP is able to estimate the viability of a plan according to several social constraints, reaching from undesirable states to effort balancing and abstraction legibility. An extension of HATP is presented by *Alili et al.* [4]. The work introduces a supervision layer that refines tasks based on the current context in an incremental fashion and monitors human behaviour, *i.e.* whether a human user commits to a task assignment or not. This allows the system to recognise in which context a human user accepts or declines a specific task. Here, the work lacks detail on the usage of such information.

Alami et al. [3] present a way to adjust the planning procedures to different types of humans using so called InterActionAgents (IAA) as a representation of humans interacting with the system similar to proxies. The author further discuss a concept for a framework using the information provided by the IAAs to produce legible behaviour.[1]

Rosenthal et al. [23] emphasise how human-agent cooperation enables a team to accomplish tasks that the team members cannot fulfil on their own. The work makes extensive use of plan changes and retasking and also introduces ways to replan in situations where individual agent capabilities are outperformed. Models of human behaviour are not used.

The work of *Kirsch et al.* [16] uses models of human abilities to predict human behaviour and reactions and is in consequence able to produce plans that are socially acceptable. Furthermore, the authors state that these models are adjusted through a learning cycle. Nevertheless, the strengths of the system are not the planning techniques or the learning algorithms but the concept of combining two frameworks to facilitate joint human-agent activities.

5 Discussion

As mentioned above several of the presented requirements can be imported 'out of the box' from related work. Nevertheless, we are able to identify subjects that are merely covered by state-of-the-art solutions. Table 1 classifies the introduced approaches in comparison to our own study named the HPLAN [2] (Human-Plan) project. The table includes the ability of the solutions to monitor the execution phase and to evaluate the viability of plan changes, which is subsumed using the term manage state. Furthermore, we evaluated the approaches abilities to detect failures during execution and to replan if necessary including retasking of the planned actions. The ability to recruit more capable agents to perform the replanning was not found in any of the approaches. One reason might be a missing exchange between research in e.g., mixed-initiative planning [7] and research in planning for joint human-agent activities. As the HPLAN projects focuses interpredictability as research goal, we further evaluated whether the approaches take representation of humans into account. Here, we distinguish social constraints, behavioural models, intention models and whether these models are static or introduce some certain kind of learning.

Table 1 emphasises that technical requirements such as manage the current state, replan and the detection of failures are satisfied by most of the approaches. This is typically done by implementing the 3-layer architecture of dynamic planning components and establishing a life-cycle of planning, executing and monitoring. Additionally, we can conclude that there exist some work related to

[1] *Kirsch et al.* [17] gives an introduction to the term 'legible behaviour' in the context of joint human-agent activities.

Table 1. A classification of human-aware planning approaches. The table entries reads as follow: + the approach supports this feature, o the approach does not fully support this feature but it supports it in some (weak) way (e.g. extension), – the approach does not support this feature.

	[20]	[4]	[16]	[8]	[23]	[3]	HPLAN
Manage State	+	+	o	+	+	o	+
Replanning	+	+	o	+	+	o	+
Failure Detection	+	o	o	o	+	o	+
Social Constraints	+	+	o	o	–	+	–
Behavioural Models	o	o	+	o	o	–	+
Intention Models	o	o	o	+	–	–	–
Learning	–	–	o	–	–	–	+

the integration of specific information about humans. For example, *Montreuil et al.* [20] presents a solution that integrates social constraints into the actual planning process by assigning a social score to evaluate plans. This social score can comprise several dimensions. In contrast to social constraints, the integration of information about behaviour (e.g., personality traits) and abilities are not as well advanced. Even so, the different authors involved highlight the need for a novel representation of humans [8,16]. The conceptual work of *Kirsch et al.* [16] presents a first approach to integrate such representations, which can be also adapted during runtime using learning techniques. The concept of InterActionAgents introduced by *Alami et al.* [3] presents a way to encapsulate different information about individuals and is a natural way when developing agent-oriented.

6 Conclusion

This article provided an overview about requirements for human-aware planning and available human-aware planning approaches. In particular, we focussed on the challenges associated with joint human-agent activities. It was argued that the decision making process of artificial agents can benefit when anticipating impending actions of humans. This requirement, which is also called interpredictability, implies shared knowledge between team-members developed through experience gained during the actual teamwork [6]. Here, a novel representation about human behaviour, abilities, habits, social rules and intentions is required. These representations can be derived, e.g., from social and psychological studies and should be adapted during the interaction with humans. Although the related work fulfils several of the presented requirements, the integration of such representations is mainly missing. It was shown that major advancement were made integrating social constraints into the planning process, that the integration of behavioural and intention models is a barely sufficient explored and that the online adaptation during the interaction is nearly non-existent nowadays. In future work, we want address these issues in the HPLAN project concatenating contemporary planning techniques with a human behavioural model and reinforcement learning techniques.

References

1. Ahrndt, S.: Improving human-aware planning. In: Klusch, M., Thimm, M., Paprzycki, M. (eds.) MATES 2013. LNCS, vol. 8076, pp. 400–403. Springer, Heidelberg (2013)
2. Ahrndt, S., Ebert, P., Fähndrich, J., Albayrak, S.: HPLAN: Facilitating the implementation of joint human-agent activities. In: Demazeau, Y., Zambonelli, F., Corchado, J.M., Bajo, J. (eds.) PAAMS 2014. LNCS, vol. 8473, pp. 1–12. Springer, Heidelberg (2014)
3. Alami, R., Clodic, A., Montreuil, V., Sisbot, E.A., Chatila, R.: Task planning for human-robot interaction. In: Bailly, G., Crowley, J.L., Privat, G. (eds.) Proc. of the sOc-EUSAI 2005, pp. 81–85. ACM, New York (2005)
4. Alili, S., Warnier, M., Ali, M., Alami, R.: Planning and plan-execution for human-robot cooperative task achievement. In: Proc. of the 19th ICAPS, pp. 1–6 (2009)
5. Allen, J., Ferguson, G.: Human-machine collaborative planning. In: Proceedings of the Third International NASA Workshop on Planning and Scheduling in Space, pp. 1–10 (2002)
6. Bradshaw, J.M., et al.: From tools to teammates: Joint activity in human-agent-robot teams. In: Kurosu, M. (ed.) HCD 2009. LNCS, vol. 5619, pp. 935–944. Springer, Heidelberg (2009), `http://dx.doi.org/10.1007/978-3-642-02806-9_107`
7. Burstein, M.H., McDermott, D.V.: Issues in the development of human-computer mixed-initiative planning systems. In: Gorayska, B., Meyer, J.L. (eds.) Cognitive Technology: In Search of a Humane Interface, ch. 17, pp. 285–304. Elsever Science B.V. (1996)
8. Cirillo, M., Karlsson, L., Saffiotti, A.: Human-aware task planning: An application to mobile robots. ACM Trans. Intell. Syst. Technol. 1(2), 1–26 (2010)
9. Cirillo, M.: Planning in Inhabited Environments: Human-Aware Task Planning and Activity Recognition. Ph.D. thesis, Örebro University (November 2010), `http://www.publications.oru.se/4DCGI/dropacta?r=23850&t=t&id=510`
10. Clark, H.H.: Using Language. Cambridge Univ. Press (1996)
11. Fähndrich, J., Ahrndt, S., Albayrak, S.: Self-explaining agents. Jurnal Teknologi (Science & Engineering) 63(3), 53–64 (2013)
12. Ghallab, M., Nau, D., Traverso, P.: Automated Planning: Theory & Practice. Morgan Kaufmann (2004)
13. Wooldridge, M.: An Introduction to Multi-Agent Systems, 2nd edn. Wiley Publishing (2009)
14. Jenkins, O.C., González, G., Loper, M.M.: Tracking human motion and actions for interactive robots. In: Proceedings of the ACM/IEEE International Conference on Human-Robot Interaction, HRI 2007, pp. 365–372. ACM, New York (2007), `http://doi.acm.org/10.1145/1228716.1228765`
15. Kephart, J.O., Chess, D.M.: The vision of autonomic computing. Computer 36(1), 41–50 (2003)
16. Kirsch, A., Kruse, T., Mösenlechner, L.: An integrated planning and learning framework for human-robot interaction. In: Proc. of the 19th ICAPS, pp. 1–6 (2009)

17. Kirsch, A., Kruse, T., Sisbot, E.A., et al.: Plan-based control of joint human-robot activities. KI – Künstliche Intelligenz 24(3), 223–231 (2010)
18. Klein, G., Feltovich, P.J., Bradshaw, J.M., Woods, D.D.: Common ground and coordination in joint activitiy. In: Rouse, W.B., Boff, K.R. (eds.) Organizational Simmulation, ch. 6, pp. 139–184. John Wiley & Sons (2005)
19. Klein, G., Woods, D.D., Bradshaw, J.M., Hoffmann, R.R., Feltovich, P.J.: Ten challenges for making automation a 'team player' in joint human-agent activity. Human-Centered Computing 19(6), 91–95 (2004)
20. Montreuil, V., Clodic, A., Alami, R.: Planning human centered robot activities. In: IEEE SMC, pp. 2618–2623 (2007)
21. Müller-Schloer, C., Schmeck, H., Ungerer, T. (eds.): Organic Computing – A Paradigm Shift for Complex Systems, vol. 1. Springer (2011)
22. Neef, M., van Rijn, M., Keus, D., Marck, J.-W.: Organizing smart networks and humans into augmented teams. In: Harris, D. (ed.) EPCE 2009. LNCS, vol. 5639, pp. 213–222. Springer, Heidelberg (2009)
23. Rosenthal, S., Biswas, J., Veloso, M.: An effective personal mobile robot agent through symbiotic human-robot interaction. In: Proceedings 9th Autonomous Agents and MultiAgent Systems (AAMAS), vol. 1, pp. 915–922 (May 2010)
24. Salvit, J., Sklar, E.: Modulating agent behavior using human personality type. In: Proceedings of the Workshop on Human-Agent Interaction Design and Models (HAIDM) at Autonomous Agents and MultiAgent Systems (AAMAS), pp. 145–160 (2012)
25. Simon, H.A.: The Sciences of the Artifical, 3rd edn. MIT Press (October 1996)
26. Sycara, K.P.: Multiagent systems. AI Magazine 19(2), 79–92 (1998), `http://www.aaai.org/ojs/index.php/aimagazine/article/view/1370`
27. Talamadupula, K., Kambhampti, S., Schermerhorn, P., Benton, J., Scheutz, M.: Planning in human-robot teaming. In: Proceedings of the 21th International Conference on Automated Planning and Scheduling (ICAPS 2011), pp. 82–89 (2011)
28. Wooldridge, M.: An Introduction to Multi-Agent Systems, 2nd edn. Wiley Publishing (2009)

Autonomous Agents in Dynamic Environment: A Necessary Volatility of the Knowledge

Emmanuel Adam, Emmanuelle Grislin, and René Mandiau

Lille Nord de France, F-59000 Lille, France
UVHC, LAMIH, F-59313 Valenciennes, France
CNRS, UMR 8201, F-59313 Valenciennes, France
firstname.name@univ-valenciennes.fr

Abstract. Completely autonomous vehicles should allow to decrease greatly the number of accident victims, and should allow gains in terms of performance and economy. Interactions among the different vehicles allowing them to choose the best path, the best behaviour is one of the main challenges. We propose in this paper a model of volatile knowledge dedicated to mobile agents on a traffic network. This model of knowledge and the principles of interactions allow to propagate new knowledge with a limited number of messages. For that, a degradation coefficient of the knowledge is proposed. The principles have been validated by a simulation with software agents, and by a real application on mobile robots acting like autonomous vehicles.

1 Introduction

Developing models of the interaction among the different vehicles is one of the main challenges [1] to optimize traffic flow with autonomous vehicles. Traditionally, two approaches allow management policies for scheduling vehicles flows: centralized approaches and distributed approaches. A means to bypass the limitations of centralized approaches (i.e. lack of flexibility, difficulty to adapt to changes) is to decentralize the traffic simulation ([2] [3]); in this context agent-based approaches seem to be the most appropriate because of their modeling advantages ([4] [5]).

The driving tasks are classified in three groups according to level of cognition [6]: Strategical (search of the best roads to reach a destination for example), Tactical (for example, tasks allowing to find the local short path to reach the other side of an intersection) and Operational (reactive actions to drive the vehicle according to the local perception).

The switch from the tactical level to the strategical level can be activated by the unability to reach the objective by using the available tactical tasks.

In this paper, we will focus on a strategical task which is 'Itinerary Planning' according to knowledge about the network load, and, in the implementation inside mobile robots, we focus on operational tasks regarding the control of the trajectory. As a traffic network is a dynamic environment where roads (or rails) can be blocked or prone to slowing down, the knowledge must represent these dynamics, this volatility of the perceived information. Our objective is not to build a new protocol of knowledge

J.B. Perez et al. (eds.), *Trends in Practical Applications of Heterogeneous Multi-agent Systems. The PAAMS Collection*, Advances in Intelligent Systems and Computing 293,
DOI: 10.1007/978-3-319-07476-4_13,

propagation, some protocols exist already, but to propose a management of dynamic knowledge, created from the perception of mobile agents.

The next section of the paper presents the model of volatile knowledge we propose and use to simulate knowledge sharing between drivers (human or not). Section 3 presents the simulator of road traffic that we use to validate our model, and section 4 presents the application of knowledge sharing between mobile robots interacting with wireless connection. The last section draws our conclusions and gives some perspectives for future research.

2 Volatile Knowledge for Mobile Agent Management

Communication of knowledge implies classically to take care of the confidence about this knowledge, confidence linked to the 'age' of the knowledge, or its origin, or its coherence with the owned knowledge.

A transport network is, as its name suggests it, composed of vertices and edges. In the research field turned to *ad hoc* networks [7] the vertices are called 'nodes' and represent mainly the intersections, the edges represent the roads or the railways.

Knowledge sharing between mobile agents along a network can be done: directly, by messages exchange, when agents are physically close enough to communicate; or indirectly through the environment (generally at the nodes that represent intersections some electronic devices allow the mobile agents to store/read information).

Most of the works relative to traffic management propose to control the flow of vehicles at the level of the intersections/nodes (see for example [8], [1]). Our objectives is to give the highest autonomy to the vehicles, so we made the choice to give to the node/intersection managers only the responsibility of the tactical behaviour of the autonomous vehicles that have to pass a crossroad. This differs from the other approaches that give to the intersection managers the control of the vehicles at the operational level, and where the strategical control is centralized. Thus, we propose an architecture where non-mobile agents are located at the nodes and communicate with mobile agents which are close enough to receive messages.

Elements of Volatile Knowledge Model. A knowledge is a partial view of the environment or of the other agents, namely for a given object o of the environment (the traffic network for example); it is (generally) an incomplete copy of it, so a representation of o with missing attributes and methods.

We define a knowledge κ_o^a (cf. def. 1) on an object o for an agent a by: o'_a, a partial view of o from a; $date_{\kappa_o}$, the date when the knowledge has been created or updated (by a or by another agent if the knowledge has been received); $builderAgent_{\kappa_o}$, the 'builder' of the knowledge (name of the agent that has created/updated the knowledge from its perception); $senderAgent_{\kappa_o}$, the 'sender' of the knowledge (name of the age nt that could have sent the knowledge to a); $conf_{\kappa_o} \in [0,1]$ the confidence that a has on κ_o^a; $deg_{\kappa_o} \in [0,1]$ the percentage of confidence degradation applied at each 'step';

$threshold_{\kappa_o} \in [0,1[$ the threshold under which the knowledge is no more considered (and has to be removed); $shareable_{\kappa_o^a}$, the fact that the knowledge is shareable or not by a.

$$\kappa_o^a = \begin{pmatrix} o'_a, date_{\kappa_o^a}, builderAgent_{\kappa_o^a}, senderAgent_{\kappa_o^a}, \\ conf_{\kappa_o^a}, deg_{\kappa_o^a}, threshold_{\kappa_o^a}, shareable_{\kappa_o^a} \end{pmatrix} \quad (1)$$

In the context of mobility on road or railway networks, these networks are represented by weighted graphs, the knowledge, in our model concerns: the existence or not of an obstacle on the roads; and the road speed limit (the weight of the edge): each driver agent (mobile agent) stores, updates the speed limit allowed and really practical on a road when it passes this one (in fact, the speed limit is updated in the knowledge if it differs of at least 20% (value arbitrary chosen) of the speed limit already stored).

Confidence and Volatility. In a dynamic environment, it is necessary to allow an automatic update, a cleaning of the outdated or invalidated beliefs.

In our model, at each step, each passage in the life cycle of an agent (perception-cognition-action) or at each 'tick' given by a simulator, the confidence on a knowledge is degraded: $conf_{\kappa_o^a} \leftarrow conf_{\kappa_o^a} \times (1 - deg_{\kappa_o^a})$ [1].

A knowledge κ_o^a can be perennial ($deg_{\kappa_o^a} = 0$) or volatile ($deg_{\kappa_o^a} > 0$) (we note $\overline{\kappa_{o_0}^a}$ a perennial knowledge). For an agent a, the first knowledge received, perceived or given about an object is considered as perennial (for example, a driver agent (mobile agent) starts with the road map, knowing the speed limits of the road; i.e. the weights of the edges). All the other knowledge relative to the same object will be considered by a as volatile .

If $\left(\overline{\kappa_{o_0}^a}, \kappa_{o_1}^a, \ldots, \kappa_{o_n}^a\right)$ is a list of knowledge that a owns on an object o, we consider that the first knowledge, $\overline{\kappa_{o_0}^a}$, is perennial ($deg_{\overline{\kappa_o^a}} = 0$).
Initially, $conf_{\overline{\kappa_{o_0}^a}} = 1$, but when volatile knowledge about o are added, this confidence is degraded; we propose that: $conf_{\overline{\kappa_{o_0}^a}} = (1 - \max_{i=1..n}(conf_{\kappa_{o_i}^a}))$. If more than one knowledge exists relatively to a same object, it is necessary to normalize the confidence values: let L be the sum of the confidences of the knowledge that owns a about the object o, $L = (\sum_{i=0}^{n} conf_{\kappa_{o_i}^a})$, for each confidence, we have: $conf_{\kappa_o^a} \leftarrow \frac{conf_{\kappa_o^a}}{L}$.

Example of knowledge adding and impact on the confidence level.
For example, let agent a be an agent whose knowledge at a given time about an object ob are $(\overline{k_0}, k_1)$, with $conf_{k_1} = 0.4$, that implies that $conf_{\overline{k_0}} = 0.6$. Next, a received a new knowledge about ob (k_2), with $conf_{k_2} = 0.7$. That implies firstly that $conf_{\overline{k_0}} = 0.3$, and next, with the normalisation, that $conf_{\overline{k_0}} \approx 0.2$, $conf_{k_1} \approx 0.3$, $conf_{k_2} \approx 0.5$.
When a receives an information with a confidence of 90% which it stores in k_3, the confidences are modified so that: $conf_{\overline{k_0}} \approx 0.05$, $conf_{k_1} \approx 0.2$, $conf_{k_2} \approx 0.3$ and $conf_{k_3} \approx 0.45$. Figure 1 represents the distribution of the confidences.

[1] Or if $conf_{\kappa_o^a}^{t_0}$ is the confidence at $time = 0$, confidence $conf_{\kappa_o^a}^{t_n}$ at $time = n$ is $conf_{\kappa_o^a}^{t_n} = conf_{\kappa_o^a}^{t_0} \times (1 - deg_{\kappa_o^a})^n$

When the confidence on a knowledge goes under the threshold, the knowledge is removed from the list of current knowledge (named *knowledgeList*) and is put in a list of 'doubtful' knowledge (named *knowledgeToCheck*).

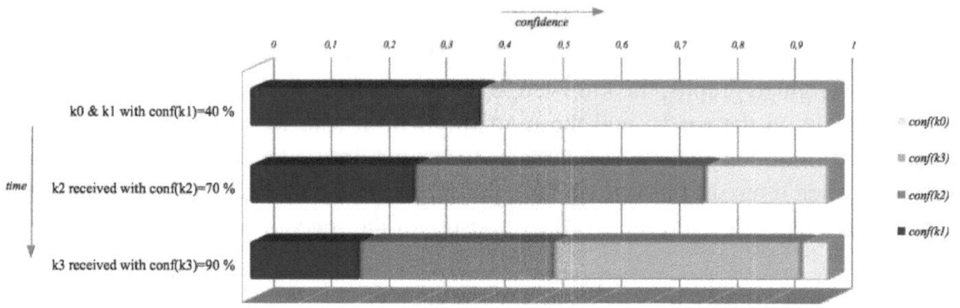

Fig. 1. Example of knowledge addition relative to a same object

A doubtful knowledge is restored and put back in the *knowledgeList* if the agent perceived directly that its information is correct, or if it receives a more recent version of the same information by another agent b ($date_{\kappa^b_o} > date_{\kappa^a_o}$).

Confidence Evolution. It occurs that different knowledge about a same object co-exist in the belief of an agent a. To choose which knowledge, which value of an object, that will be taken into account to evaluate a strategy, the agent a uses a fitness proportionate selection (i.e. roulette wheel selection) with the following distribution:

- let $\left(\overline{\kappa^a_{o_0}}, \kappa^a_{o_1}, \ldots, \kappa^a_{o_n}\right)$ be a list of knowledge that a owns on an object o,
- let x be a number randomly chosen in $[0,1[$. x is situated in an interval that defines the knowledge to select.
- the interval corresponding to the knowledge $\kappa^a_{o_j}$ with $j > 0$ is:

$$\kappa^a_{o_j} ::= \left[\Sigma_{i=1}^{j-1} conf_{\kappa^a_{o_i}}, \Sigma_{i=1}^{j} conf_{\kappa^a_{o_i}}\right[$$

- the interval corresponding to the perennial knowledge $\overline{\kappa^a_{o_0}}$ is:

$$\overline{\kappa^a_{o_0}} ::= \left[\Sigma_{i=1}^{n} conf_{\kappa^a_{o_i}}, 1\right[$$

Example with traffic management.
If we come back to the previous example, the agent a owns the knowledge $(\overline{k_0}, k_1, k_2, k_3)$ about the same object ob. Let ob be a road object and $\overline{k_0}$ the initial and perennial information that the speed limit on ob is of $90km/h$, with a confidence of 5%. k_1 comes from a perception, k_2 and k_3 have been received from other agents.
k_1 informs that the speed limits have been detected as being of $20km/h$ at $date =$ 30, k_2 informs that the speed limits have been detected as being of $50km/h$ at $date = 90$, k_3 informs that the speed limits have been detected as being of $75km/h$ at $date = 110$.
At $time = 110$ $Min.$, the distribution of confidence follows those specified in the previous example: $conf_{\overline{k_0}} \approx 0.05$, $conf_{k_1} \approx 0.2$, $conf_{k_2} \approx 0.3$ and $conf_{k_3} \approx 0.45$.
If we suppose that the agent a sets the degradation coefficient of k_1, k_2 and k_3 to 0.2, and that it sets the threshold under which a knowledge is no more considered at 0.07, the evolution of the confidences is given Figure 2.
At time = 110 Min., the agent has relatively equal probability to select k_1 or k_2, less chance to choose k_1 and only few chances to consider $\overline{k_0}$ before taking a decision. With the time, only the perennial knowledge $(\overline{k_0})$ stays in the memory of the agent (when time = 200 Min. in this example).

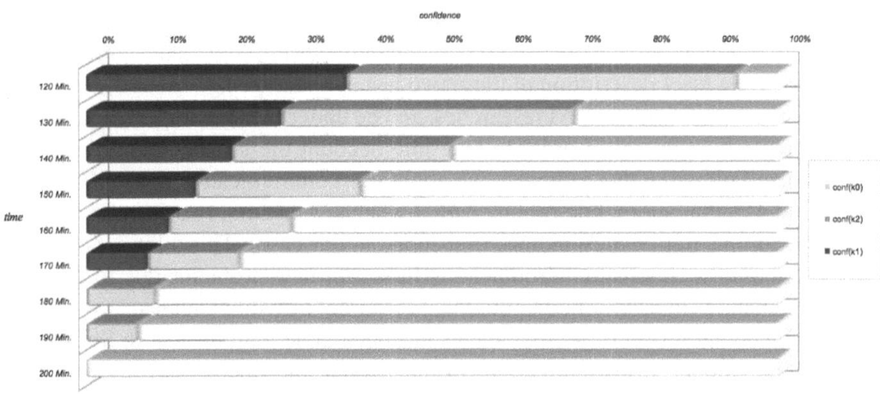

Fig. 2. Example of knowledge predominance evolution

This principle of knowledge management is dedicated to agents evolving in a dynamic environment, subject to modifications; it allows an agent to take into account different observations about a same object, in order to take a decision.

During an action, the agent is able to check if an information is still valid or not, and modify its knowledge list consequently.

In the previous example, if a perceives next that the speed limit of the road is $75km/h$, the confidence of k_3 is set to 1 and the knowledge $k1, k2$ are removed. k_0 persists, with a confidence of 0, that will grow with the degradation of $conf_{k_3}$.

Adjustment of the Degradation Coefficient. The model of knowledge allows to adjust, more than learn, the value of the degradation coefficient. Indeed, the more some knowledge about a same object o are added to the list, the more the degradation coefficient of the knowledge about o increases.

When an agent perceives or receives an information similar to one in its knowledge list, or in its list of 'doubtful' knowledge, the confidence is restored if needed (if the new confidence perceived $conf^{t_{n'}}$ is 'significantly' different than the computed confidence $conf^{t_n}$), and the degradation coefficient is adapted:

- if $conf^{t_0}$ was the knowledge confidence at $time = 0$,
- if $conf^{t_{n'}}$ is the new confidence of the knowledge perceived or received at $time = n$, with $conf^{t_{n'}} > conf^{t_n}$ ($conf^{t_n}$ being the confidence computed by the agent),
- as we have $conf^{t_{n'}}_{\kappa^a_o} = conf^{t_0}_{\kappa^a_o} \times (1 - deg'_{\kappa^a_o})^n$,

  the new degradation coefficient to apply is $deg'_{\kappa^a_o} = \left(1 - \sqrt[n]{\frac{conf^{t_{n'}}}{conf^{t_0}}}\right)$

3 Applications of Volatile Knowledge: Management of Autonomous Vehicles

We have applied our model on different applications, like traffic road simulation, management of autonomous vehicles and flexible manufacturing cells management. In this paper, we give a brief description of these two last applications.

In [9], the author criticizes AI researches, and he underlines the importance of working in a real environment in order to take into account the sensory-motor aspects of intelligence. Following this criticism, we apply our communication model to wifibots (cf. Figure 4)[2], that are mobile robots, with 4 wheels, a camera, two InfraRed sensors, and embedding a light OS (Windows Embedded in our case). This allows us to examine the effect of a concrete environment and identify new problems due to the really distributed nature (in terms of computational resource) of wifibot agents.

The conceptual model of wifibot agents has been done with respect to Strate-gical-Tactical-Operational levels distinction. In this model, the operational level takes the responsibility of commanding the robot engines like forward, backward, go left, go right, etc. After sending a command, the engine gives a response including the information of current speed, voltage, odometric information, infrared sensors informations (IR sensors), etc.

The tactical level accomplishes the composite behaviors like intersection crossing, U-turn, etc. Our wifibot is equipped with a camera and a perception module linked to the tactical level that processes images taken by camera in order to identify the behavioral triggers (e.g., an identified intersection triggers the intersection crossing behavior).

The strategical level executes itinerary control and change. Agents have a destination point and a best itinerary to reach this destination point. In case of an unpredictable event like an incident or a blocked way somehow, agents deliberate for finding an alternative itinerary. Our communication model is applied at this stage, and allows to exchange among agents the information about the incident.

[2] See the web site relative to the WifiBot: `http://www.wifibot.com/`

In the scenario, illustrated Figure 3, robots have to make some loops between the 'start' and 'objective' points, with a shortest path[3].

Fig. 3. Scenario for the WifiBots

Fig. 4. WifiBot

Initially the robots make loops through the nodes 1, 2, 5, 6. Some obstacles are added manually. In the example proposed Figure 3, the robot at the node 1 detects the obstacle on the edge [2-5] and makes a U-turn to take the edge [3-4]. The second robot, informed by the first robot, decides also to use the same path [1-3-4-6]. When the first robot detects the obstacle on the edge [3-4], it sends the information to the second robot; and stays on this edge, because it knows that the other robot will test the edge [2-5]: the two wifibots have the same behaviours and know it; information about [2-5] being older than about [3-4], there is more probability that the second robot tests the first edge. If the obstacles are not removed, the robots stay on their respective edge until one of the obstacle is removed. Then, the released robot informs the other of the opening of the edge and continues its tasks. The other robot make a U-turn to take the released edge. If this one is the edge [3-4], the two robots make the 'big loops' [1-3-4-6] until one robot degrades enough of the confidence of the edge [2-5] to remove it and test this edge.

Other scenarios have been implemented, with one[4], two and three robots to validate our model of knowledge.

4 Conclusion

In order to allow the propagation of knowledge between mobile agents in a dynamic environment, a model of knowledge and a model of communication have been presented. This model, simple, can be be easily embedded in light devices. It allows also to propagate knowledge about disruptions, and return to a normal situation, in a distributed way without any coordinator.

[3] A video can be read here: `http://www.youtube.com/watch?v=ifdCp76BKnM`

[4] A video can be watched: `http://www.youtube.com/watch?v=gMiYeEH_KLU`

In the results presented in this paper, we make the assumption that all the agents are cooperative, and no defective; that is to say that they cannot send wrong knowledge, voluntary, or not (if a sensor has a dysfunction).

The trust on knowledge depends only on the date from which the knowledge has been updated or created. We plan to introduce the notion of trust that depends of the sender; and thus, the notion of suspicious agent. When the suspicious potential of an agent reaches a threshold, this agent is considered as suspect and its communications are no more considered, or with a low level of confidence. This information is communicated to other agents, these ones can decide to put in quarantine the faulty agent, like in [10] for example.

Acknowledgement. This research was financed by the International Campus on Safety and Inter-modality in Transportation, the Nord/Pas-de-Calais Region, the European Community, the French Regional Delegation for Research and Technology, and the French National Centre for Scientific Research. We are grateful for the support of these institutions.

The authors thank also Utku Ketenci for his contribution in this work.

References

1. Dresner, K., Stone, P.: Mitigating catastrophic failure at intersections of autonomous vehicles. In: AAMAS Workshop on Agents in Traffic and Transportation, Portugal, pp. 78–85 (May 2008)
2. Mandiau, R., Champion, A., Auberlet, J.M., Espié, S., Kolski, C.: Behaviour based on decision matrices for a coordination between agents in a urban traffic simulation. Appl. Intell. 28(2), 121–138 (2008)
3. Ruskin, H.J., Wang, R.: Modelling traffic flow at an urban unsignalised intersection. In: Sloot, P.M.A., Tan, C.J.K., Dongarra, J., Hoekstra, A.G. (eds.) ICCS 2002, Part I. LNCS, vol. 2329, pp. 381–390. Springer, Heidelberg (2002)
4. Klügl, F.: Agent-based simulation engineering. Habilitation thesis (July 2009)
5. Macal, C.M., North, M.J.: Agent-based modeling and simulation: ABMS examples. In: Proceedings of the 40th Conference on Winter Simulation, WSC 2008, pp. 101–112 (2008)
6. Allen, T.M., Lunenfeld, H., Alexander, G.J.: Driver information needs. Highway Research Board 36, 102–115 (1971)
7. Lochert, C., Hartenstein, H., Tian, J., Fussler, H., Hermann, D., Mauve, M.: A routing strategy for vehicular ad hoc networks in city environments. In: Proceedings of the IEEE Intelligent Vehicles Symposium, pp. 156–161 (2003)
8. Bazzan, A.L.: A distributed approach for coordination of traffic signal agents. Autonomous Agents and Multi-Agent Systems 10, 131–164 (2005)
9. Brooks, R.: Intelligence without representation. Artificial Intelligence 47, 139–159 (1991)
10. Vercouter, L., Jamont, J.-P.: Lightweight trusted routing for wireless sensor networks. In: Demazeau, Y., Pechoucek, M., Corchado, J.M., Pérez, J.B. (eds.) PAAMS 2011. AISC, vol. 88, pp. 87–96. Springer, Heidelberg (2011)

RETRACTED CHAPTER: Indoor Location System for Security Guards in Subway Stations

Juan Francisco De Paz [1], Gabriel Villarrubia[1], Javier Bajo[2], Gabriel Sirvent[3], and Tiancheng Li[4]

[1] Department of Computer Science and Automation, University of Salamanca, Plaza de la Merced, s/n, 37008, Salamanca, Spain
{gvg,fcofds}@usal.es
[2] Department of Artificial Intelligence. Faculty of Computer Science. Technical University of Madrid, Madrid, Spain
jbajo@fi.upm.es
[3] Intelligence Artificielle, University of Toulouse, 15 Rue des Lois, 31000 Toulouse France
sirventgabriel@gmail.com
[4] School of Mechatronics, Northwestern Polytechnical University, Xiniversity, China
t.c.li@mail.nwpu.edu.cn

Abstract. Indoor locating systems (RTLS), have notably advanced during recent years, becoming one of the main challenges for several research teams. The main objective of indoor locating systems is to obtain functional systems able to locate different elements in those environment where GPS (Global Positioning System) is limited. The growing use of mobile devices in the information society provides a powerful mechanism to obtain geographical data and has led to new algorithms aimed at facilitating object positioning with easonable power consumption. In this paper we propose an innovative indoor location architecture that makes use of the data provided by mobile devices to locate objects. The architecture is applied to a case study in a real environment focused on obtaining the location of security staff in the subway network in a city in the north of Spain.

Keywords: indoor locating system, Wi-Fi, MQTT.

1 Introduction

Indoor locating systems have significantly evolved during recent years [10] [13] [15] [3]. The main challenge of current research in indoor locating systems is to provide a system that can calculate the position of different resources in indoor systems with a reduced cost [1][2]. Estimating the location of a given resource constitutes the basis of the design of advanced services such as resource identification, security,

---

The original version of this chapter was retracted: The retraction note to this chapter is available at https://doi.org/10.1007/978-3-319-07476-4_22

J.B. Perez et al. (eds.), *Trends in Practical Applications of Heterogeneous Multi-agent, Systems. The PAAMS Collection*, Advances in Intelligent Systems and Computing 293, DOI: 10.1007/978-3-319-07476-4_14, © Springer International Publishing Switzerland 2014, corrected publication 2024

recommendation systems, human behaviour analyzers, etc. [28][30]. The majority of these services are deployed in indoor environments such as hospitals, subway stations, shopping malls and so on, characterized by a weak or inexistent GPS signal. Thus, taking into account the importance of location-based services[11][14], it is necessary to design new and effective indoor locating systems.

Nowadays, the use of Wi-Fi technology allows the development of real time indoor locating systems with a low-cost infrastructure[16][17] compared to alternative technologies such as (*Radio Frequency iDentification*) [5] or ZigBee[6]. The technological infrastructure required to provide indoor location use Wi-Fi communication, which is based on the deployment of Access Points in the environment to be controlled. These access points can be used as readers. One of the advantages of this kind of solution is that Wi-Fi systems are very common and it is usual to find Wi-Fi[20][22] access points installed in a broad variety of public places, hospitals or environments dedicated to leisure activity. The existing technology[18] can be reused to implement indoor location solutions without additional costs. Furthermore, the use of wireless technologies makes it possible to locate people by means of their smartphones or another electronic low-cost devices.

In this paper we propose an indoor location architecture based on Wi-Fi technology that provides the location of the subway security guards and determines the station where the security guard is located at any given moment. The systems calculates the position of the security guard based on his or her personal smartphone[23][12][27], which presents several limitations including connectivity problems and battery consumption. The smartphone is not only used as a tag to be located, but also as a way to comminicate with other security guards. The architecture defines a new algorithm to estimate the real positon of the users using GSM (Global System for Mobile) signals and resource-constrained devices. The algorithm uses reduced data frames and the infrastructure uses a low-latency setting, which facilitates effective communications with an external server.

The rest of the paper is organized as follows: Section 2 revises the state of the art of the situation that has motivated this research, Section 3 presents the architecture and its application to a case study. Finally Section 4 present the results obtained.

2 Background

Nowadays it is possible to find differnt algorithms and commercial systems aimed at facilitating the indoor location of a user. Table 1 presents a comparative study between the main existing systems, identifying their strengths and weaknesses.

Existing systems [19] [21] present business models with several restrictions, based on pay-per-use systems and hardware renting systems.

Table 1. Comparison Between Commercial Systems

DEVICE	THECNOLOGY	TYPE	SIZE (CM)	ACCURACY (M)	ADVANTAGES	DISADVANTAGES
HELICOMM (EZ-TRACER)	ZigBee	Beacons	6.8x4x1.7	3	Development Kit	Very Expensive
EKAHAU	WiFi	Tag	4.5x5.5x1.9	5	Through wall	Calibration Process
AEROSCOUT	WiFi-RFID-UWB	Beacons	7.4x5x1	15	Movement sensor	Calibration Process
UBISENSE	UWB 7Ghz-TDOA-AOA	Beacons	8.3x4.2x1.1	0.3	-	Without Warranty
SAPHIRE MULTISPECTRAL	TDOA 5.94Ghz	Tag	2.8x2.8x2.54	1	Only US	-
AWARE POINT	ZigBee	Tag	1.6x2.7x0.5	10	ZigBee Alliance	Very Expensive
TIMEDOMAIN	UWB	Tag	1.9x2.3x0.9	15	Health Enviro-ment	Poor Informa-tion
PLACELAB	WiFi	Beacons	1.9x2.8x1.9	1	Open Source	Microsoft

The common feature of all the systems evaluated [25] [29] (and main disadvantage) is the introduction of proprietary hardware, which does not make is possible to use a conventional smartphone to identify location. Furthermore, many existing systems require [24] [26] the use of a tag component. This item type is invasive for users and lacks functionality if a similar service can be offered by using a mobile terminal.

3 Location Architecture for Subway Environments

This section presents an arquitecture especially designed to locate human resources in indoor environments, more specifically in subway stations. The special characteristics of this environment require the design of an innovative solution. The main objective of the proposed architecture is to locate security staff in subway stations. Different alternatives have been evaluated to design the architecture. The proposed system uses GSM signals and smartphones, due to the execessive cost that the use of wi-fi tags can have. The system infrastructure, which is shown in Figure 1, is composed of two different sub-systems: central server and mobile devices. From the software point of view the architecture proposes two elements: the location engine and the location system, both of which are explained in the following sub-sections.

3.1 Communication System

This system interconnects the mobile devices with a central server, which is in charge of calculating positions from the information acquired by the mobile devices.

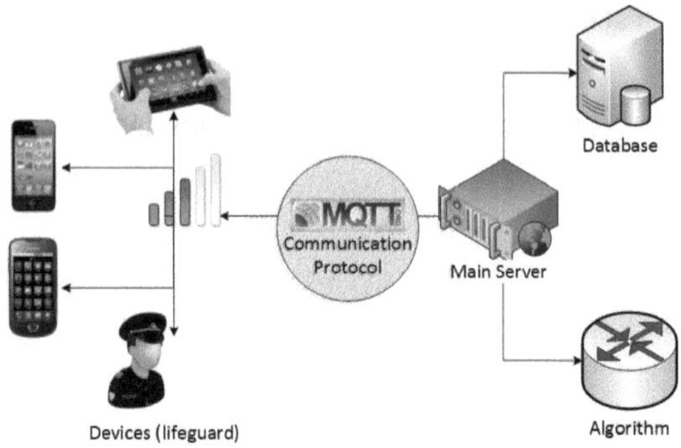

Fig. 1. Proposed architecture

***Mobile Devices*:** An Android application is installed in each of the mobile devices executed as a service. The application periodically scans the wireless networks detected in the surroundings and sends the information collected to the central server. The data frame format contains the information of the detected networks, the time of the scan, the mobile device identifier and the percentage of battery of the device.

The Wi-Fi networks detected by the mobile devices correspond to the access points installed in the subway stations. The Wi-Fi infrastructure for each of the subway stations consists of two Access points, model UBNT PICOSTATION M2H, on either side of the platform. The stations have a length of 100m while the Access points can cover 150 m.

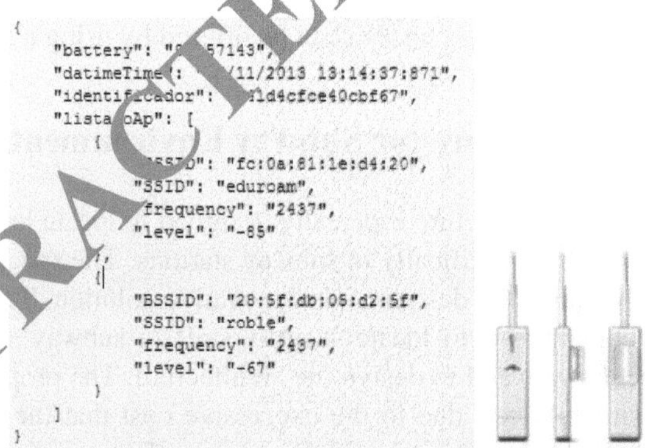

Fig. 2. Mobile Datagram Example. Example of Beacon

Central Server: The mobile devices capture the data which are sent to a central server using an innovative communication system called MQTT (MQ Telemetry Transport). The use of a communication protocol as Machine-to-Machine (M2M) allows data transmission in high latency networks with certain constraints [31]. It is a lightweight protocol that allows a publication-subscription mechanism, which is very useful for a low consumption bandwidth. In the architecture presented in this paper, the subway stations are not equipped with a high speed bandwidth and the communication between the mobile devices and the central server is implemented by means of GPRS (General Packet Radio Service).

One of the most innovative aspects of the proposed architecture [32] is the integration within a MQTT protocol for Information exchange. This provides high efficiency compared to the existing commercial systems that use SOA (Service Oriented Architecture). The comparison shown in Table 2 shows underscores the advantages of using a MQTT protocol instead of a traditional client/server architecture in terms of battery consumption. Table 2 presents a comparison between MQTT and HTTPS messages for a total of 1024 communications [7].

Table 2. Comparative Battery saving

	3G		WiFi	
	HTTPS	MQTT	HTTPS	MQTT
% Battery / Hour	18.43%	16.13%	3.45%	[illegible]
Messages / Hour	1708	160278	[illegible]	263314
% Battery / Message	0.01709	0.00010	0.00095	0.00002
Messages Received	240 / 1024	[illegible]	524 / 1024	1024 / 1024

As can be seen in Table 2, the use of MQTT can be justified in terms of battery savings alone.

3.2 Location System

The location algorithm presented in this paper uses a signpost technique to calculate the station where a security guard is located [4]. Signpost is characterized by its simplicity and a low computational cost [8]. The location of each of the security guards is estimated detecting the more powerful signal as shown in Figure 4.

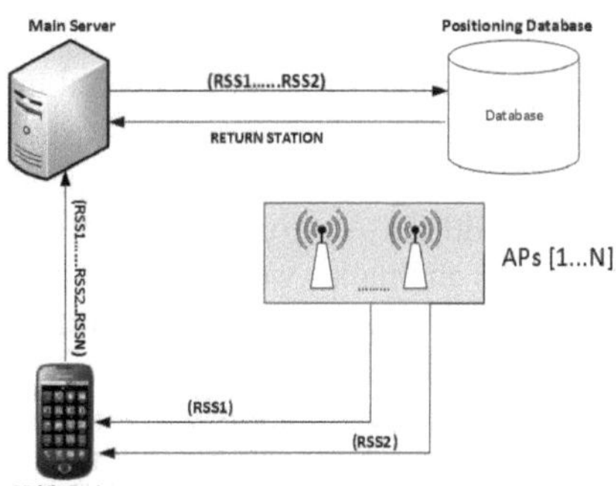

Fig. 3. Wifi Architecture

As seen in Figure 5, the 12 subway stations in the case study used to evaluate the architecture in this paper are separated by a considerable distance. As a result, the routers do not interfere with each other, and can estimate the station where there is a guard with a rate of 100%.

Station	1	2	3	4	5	6	7	8	9	10	11	12
1	0	250	296	312	314	288	287	295	293	300	311	299
2		0	298	288	302	287	298	296	287	301	280	291
3			0	289	308	319	317	285	317	286	320	302
4				0	297	298	297	292	285	315	308	320
5					0	297	298	285	296	289	302	294
6						0	296	305	317	299	306	316
7							0	268	286	316	298	310
8								0	269	281	293	296
9									0	280	312	316
10										0	254	295
11											0	281
12												0

Fig. 4. Distances between stations in meters

To make the location process work correctly, it is necessary to link the metro stations with the Wi-Fi beacons that have been deployed in the case study. This relationship is implemented using a relational database and the MAC addresses of the associated access points are stored for each station, as shown in Figure 6.

STATION	LATITUDE	LONGITUDE	XPOS	YPOS	ID	XLABEL	YLABEL	TIPE
STATION 1	43.401.959	-2.946.509	[illegible]	55	PLT	-60	-15	2

Station	MAC
STATION 1	dc:9f:db:9a:9f:27
STATION 2	dc:9f:db:9a:9f:c8

Fig. 5 Relationship between stations and routers

The central server stores the location data of the security guards in a database. These data are used to generate analysis and extract knowledge.

4 Results and Conclusions

The architecture presented in Section 3 has been installed in a subway network in the city of Bilbao in the north of Spain. The system infrastructure included 42 access points in 21 stations, and the security guards were equipped with 12 mobile devices, model GT-i9502 MTK6589, with an Android 4.2 operating system. The system was tested during one month and no anomalies were detected during the evaluation.

The central server was used by a supervisor to control the location of the security guards. To facilitate this task we developed a visualization tool that presents the location of the security guards in a very intuitive way. The central server implements monitoring services that are used by the supervisor to define alarms that are triggered when a security guard deviates from his or her usual route.

Figure 7 shows a screenshot of the central server application which is installed in the central server and allows monitoring the security guards.

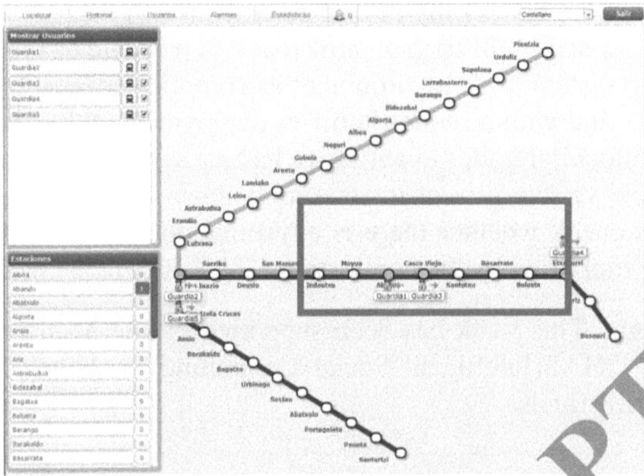

Fig. 6. Utility to view the positions

The users' location is updated in the visualization tool using Node.js technologie [9], an asynchronous platform that allows the development of scalable networks in a very quick way. The location of each security guard is asynchronously and immediately updated when a frame from a mobile device is received, without any refresh or request message required.

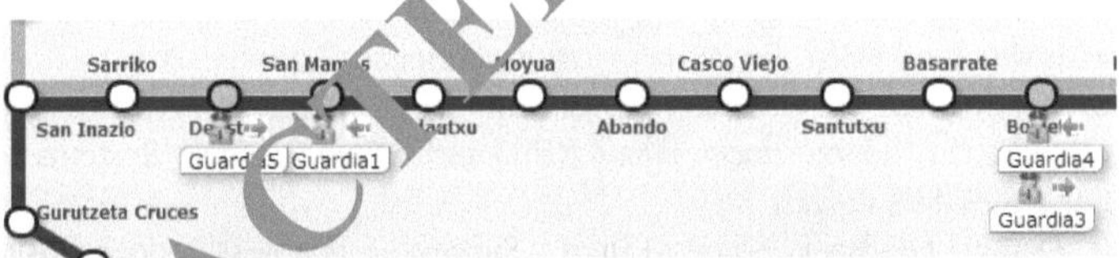

Fig. 7. Detailed View of lifeguard movements

System users have highlighted the ease of use and accuracy of the architecture. One of the most appreciated aspects among supervisors was the ability to generate multiple reports from the location data of the security guards stored in the database. For the supervisor it was of special interest to know the time spent by a security guard in a particular station, as shown in Figure 8.

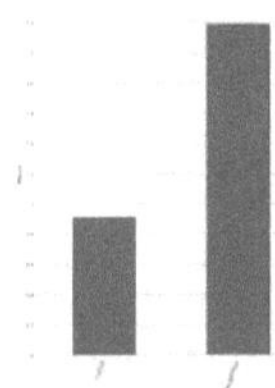

Fig. 8. Report Example

Different tests were conducted to evaluate the battery consumption by comparing the proposed system with commercial systems. The use of a protocol with low latency and the avoidance of using conventional web services, yielded an average of 27% in terms of optimizing the battery life of mobile devices.

The results presented in this paper show that it is possible to implement a functional indoor locating system in an environment as complicated as subway stations, using wireless networks and with a reduced cost of deployment. It has been shown that the use of a M2M protocol provides quantitative battery savings in mobile terminals.

Finally, we believe that it is of interest to evaluate the system in a more complex environment and check whether there is any functional limitation on the number of users that can be supported by the application. This is our next challenge.

Acknowledgments. This work has been supported by the Ministry of Economy and Competitiveness "iHAS(Intelligent Social Computing for Human-Agent Societies)" financed by FEDER funds.

References

1. Anand, P., Siva Prasad, B.V., Venkateswarlu, C.: Modeling and optimization of a pharmaceutical formulation system using radial basis function network. International Journal of Neural Systems 19(2), 127–136 (2009)
2. Carnicer, J.M.C., García-Esnaola, M.: Lagrange interpolation on conics and cubics. Comput. Aided Geom. Design. 19, 313–326 (2002)
3. Chen, Y.-C., Chiang, J.-R., Chu, H.-H., Huang, P., Tsuid, A.W.: Sensor-Assisted Wi-Fi Indoor Location System for Adapting to Environmental Dynamics (2011)
4. Huang, C.N., Chan, C.T.: ZigBee-based indoor location system by k-nearest neighbor algorithm with weighted RSSI. Procedia Computer Science 5, 58–65 (2011)
5. Tesoriero, R., Tebar, R., Gallud, J.A., Lozano, M.D., Penichet, V.M.R.: Improving location awareness in indoor spaces using RFID technology. Expert Systems with Applications 37(1), 894–898 (2010)
6. Hui, L., Darabi, H., Banerjee, P., Liu, J.: Survey of Wireless Indoor Positioning Techniques and Systems. IEEE Transactions on Systems, Man, and Cybernetics, Part C: Applications and Reviews 37(6), 1067–1080 (2007)
7. Web, http://stephendnicholas.com/archives/1217 (last visited January 14, 2014)
8. Tapia, D., Bajo, J., De Paz, J.F., Alonso, R.S.: Using Multi-Layer Perceptrons to Enhance the Performance of Indoor RTLS (2011)
9. Web, http://nodejs.org (last visited January 14, 2014)
10. Corchado, J.M., Fyfe, C.: Unsupervised neural method for temperature forecasting. Artificial Intelligence in Engineering 13(4), 351–357 (1999)
11. Fdez-Riverola, F., Corchado, J.M.: CBR based system for forecasting red tides. Knowledge-Based Systems 16(5), 321–328 (2003)
12. Tapia, D.I., Abraham, A., Corchado, J.M., Alonso, R.S.: Agents and ambient intelligence: case studies. Journal of Ambient Intelligence and Humanized Computing 1(2), 85–93 (2010)
13. Corchado, J.M., Lees, B.: Adaptation of cases for case based forecasting with neural network support. In: Soft Computing in Case Based Reasoning, pp. 293–319 (2001)
14. Corchado Rodríguez, J.M.: Redes Neuronales Artificiales: un enfoque práctico. Servicio de Publicacións da Universidade de Vigo, Vigo (2000)

15. Bajo, J., Corchado, J.M.: Evaluation and monitoring of the air-sea interaction using a CBR-Agents approach. In: Muñoz-Ávila, H., Ricci, F. (eds.) ICCBR 2005. LNCS (LNAI), vol. 3620, pp. 50–62. Springer, Heidelberg (2005)
16. Fraile, J.A., Bajo, J., Corchado, J.M., Abraham, A.: Applying wearable solutions in dependent environments. IEEE Transactions on Information Technology in Biomedicine 14(6), 1459–1467 (2011)
17. Corchado, J.M., De Paz, J.F., Rodríguez, S., Bajo, J.: Model of experts for decision support in the diagnosis of leukemia patients. Artificial Intelligence in Medicine 46(3), 179–200 (2009)
18. De Paz, J.F., Rodríguez, S., Bajo, J., Corchado, J.M.: Case-based reasoning as a decision support system for cancer diagnosis: A case study. International Journal of Hybrid Intelligent Systems 6(2), 97–110 (2009)
19. Tapia, D.I., Rodríguez, S., Bajo, J., Corchado, J.M.: FUSION@, a SOA-based multi-agent architecture. In: International Symposium on Distributed Computing and Artificial Intelligence 2008 (DCAI 2008), pp. 99–107 (2008)
20. Corchado, J.M., Aiken, J.: Hybrid artificial intelligence methods in oceanographic forecast models. IEEE Transactions on Systems, Man, and Cybernetics, Part C: Applications and Reviews 32(4), 307–313 (2002)
21. Corchado, J.M., Aiken, J., Rees, N.: Artificial intelligence models for oceanographic forecasting. Plymouth Marine Laboratory (2001)
22. Rodríguez, S., Pérez-Lancho, B., De Paz, J.F., Bajo, J., Corchado, J.M.: Ovamah: Multiagent-based adaptive virtual organizations. In: 12th International Conference on Information Fusion, FUSION 2009, pp. 990–997 (2009)
23. Tapia, D.I., De Paz, J.F., Rodríguez, S., Bajo, J., Corchado, J.M.: Multi-agent system for security control on industrial environments. International Transactions on System Science and Applications Journal 4(3), 222–226 (2008)
24. Borrajo, M.L., Baruque, B., Corchado, E., Bajo, J., Corchado, J.M.: Hybrid neural intelligent system to predict business failure in small-to-medium-size enterprises. International Journal of Neural Systems 21(04), 277–296 (2011)
25. De Paz, J.F., Rodríguez, S., Bajo, J., Corchado, J.M.: Mathematical model for dynamic case-based planning. International Journal of Computer Mathematics 86(10-11), 1719–1730 (2009)
26. Bajo, J., De Paz, J.F., Rodríguez, S., González, A.: Multi-agent system to monitor oceanic environments. Integrated Computer-Aided Engineering 17(2), 131–144 (2010)
27. Tapia, D.I., Alonso, R.S., De Paz, J.F., Corchado, J.M.: Introducing a distributed architecture for heterogeneous wireless sensor networks. In: Omatu, S., Rocha, M.P., Bravo, J., Fernández, F., Corchado, E., Bustillo, A., Corchado, J.M. (eds.) IWANN 2009, Part II. LNCS, vol. 5518, pp. 116–123. Springer, Heidelberg (2009)
28. Rodríguez, S., de Paz, Y., Bajo, J., Corchado, J.M.: Social-based planning model for multiagent systems. Expert Systems with Applications 38(10), 13005–13023 (2011)
29. Pinzón, C.I., Bajo, J., De Paz, J.F., Corchado, J.M.: S-MAS: An adaptive hierarchical distributed multi-agent architecture for blocking malicious SOAP messages within Web Services environments. Expert Systems with Applications 38(5), 5486–5499
30. Corchado, J.M., Bajo, J., De Paz, J.F., Rodríguez, S.: An execution time neural-CBR guidance assistant. Neurocomputing 72(13), 2743–2753 (2009)
31. Gómez, J., Patricio, M.A., García, J., Molina, J.M.: Communication in distributed tracking systems: an ontology-based approach to improve cooperation. Expert Systems 28(4), 288–305 (2011)
32. Pavon, J., Sansores, C., Gomez-Sanz, J.J.: Modelling and simulation of social systems with INGENIAS. International Journal of Agent-Oriented Software Engineering 2(2), 196–221 (2008)

RETRACTED CHAPTER: Mobile-Based Recommendation system in Conferences

Daniel Hernández[1], María Moreno[1], and Sigeru Omatu[2]

[1] Departamento de Informática y Automática, Universidad de Salamanca
Plaza de la Merced, s/n, 37008, Salamanca, España
{danihiglesias,mmg}@usal.es
[2] Department of Electronics, Information and Communication Engineering,
Osaka Institute of Technology, Osaka 535-8585, Japan
omatu@rsh.oit.ac.jp

Abstract. Today information is a very important asset for organizations. Obtaining and interpreting information in real time can be a major benefit in decision making. For this reason, we have developed a mobile application that uses WiFi access points to locate users who are attending conferences and exhibits. The application can generate a dataset according to the user's preferences and on-site location, and provide personalized recommendations accordingly.

1 Introduction

At present, indoor locating using WiFi technology is one of the most promising systems for obtaining accurate information about the location of objects or people inside buildings. As a result of the boom in Smartphone and mobile connections, it has become easier to locate users and objects inside a building [1]., This technology can be applied to environments such as hospitals [2], factories, buildings [3] [4], shopping malls, offices, etc ... Some of the applications and utilities to monitor real-time users in a given environment include safety, guidance, interaction with the environment or data collection, among others[8][9][10][11][12][13][14].

Most existing applications for conference and event attendees simply provide a fixed list of timetables and events for informational purposes. We intend to go one step further by studying the movements of users as they make their way around the conference, and giving the users more filtered and personalized information that can help them decide which session and events to attend sessions. Conferences and events usually last a couple of days, during which dozens of events are held simultaneously making it impossible for one person to attend all of them. Therefore, it would be exceedingly useful to have a tool that stores user preferences and assists the user in deciding, in real time, which presentation or event to attend[15][16][17][18][19][20].

This research has created a system that tracks users attending a conference, gathers their location data, and generates a profile of tastes and preferences according to the rooms that the user has visited, the amount of time spent in each one, and the articles presented in each session. In order to track conference attendees, the system scans the area using WiFi network fingerprints to determine the location of each user at any time. The information obtained can be used to analyze the user's preferences and

The original version of this chapter was retracted: The retraction note to this chapter is available at https://doi.org/10.1007/978-3-319-07476-4_22

J.B. Perez et al. (eds.), *Trends in Practical Applications of Heterogeneous Multi-agent, Systems. The PAAMS Collection*, Advances in Intelligent Systems and Computing 293,
DOI: 10.1007/978-3-319-07476-4_15, © Springer International Publishing
Switzerland 2014, corrected publication 2024

recommend other sessions that the user may wish to attend. Collaborative filtering techniques are used to make recommendations, similar to films available throughNetflix recommended mode. The article is structured as follows: Section 2 reviews the state of the art, section 3 presents the proposed system, and Section 4 includes the results and conclusions.

2 Background

In today's market there are several online payment services to build and manage applications for conferences and events. One example is guidebook (www.guidebook.com). This payment service offers a CMS system to enter and manage information about conferences and conventions that is subsequently loaded into an application for Smartphones from which attendees can display information about events, evaluate the different sessions or share it on social networks. Another similar service is bizzabo (www.bizzabo.com), another payment service that offers a platform where the user enters all the information sessions and events, which will be displayed on the mobile application.

There are several technologies for creating indoor location systems. The location system based on passive RFID (Radio Frequency IDentification) [5] uses an RFID tag that contains a receiving and transmitting antenna. The antenna transmits information when it receives a signal from the transmitter.

There are two kinds of RFID system tags: passive, which do not have a battery; and active, which does contains a battery. The sensing distance with RFID passive tags is lower than with active tags. When we work with passive tags the distance is less than one meter, while with active tags the distance increases to 100 meters. Another wireless technology is Bluetooth. These devices can be located according to the signal level of several static Bluetooth devices. While the main advantage of Bluetooth is the price, the sensing distance is low and it would be necessary to include many devices, such as a beacon, to locate the tags. The error rate is low due to the short sensing distance, which is necessary because the RSS indicator is not very precise. Some authors combine Bluetooth and WiFi in order to improve the precision of the system [6].

Still other authors use ZigBee technology. The specification was created by a consortium of businesses based on the 802.15.4 IEEE standard [7]. ZigBee technology is characterized by a low transmission rate, around 250 kbps. It makes it possible to create a low cost system, using simple protocols that can be deployed in a microcontroller with 8 bits and low energy power. Low consumption is important in order to create devices with long life batteries.

Finally, there is another technology based on the 802.11 standard. This technology calculates the position of the devices according to the levels of RSSI signal from the detected access points. To estimate the position of the tags it is possible to apply a triangulation or a fingerprint. When fingerprinting, a calibration process is carried out and we obtain a set of values for several positions. The device to be located scans the WiFi networks; the values are obtained and the fingerprint estimates the locations.

When we work with recommendation systems there are two different types of information we can use: data from the user profile or the user's rating of the items [8]. The first is used in ColdStart: the system would recommend an item according to the similarity it holds to the user profile. However, in order to do this it is necessary to calculate a measure of similarity among the users, and some attributes cannot be easily compared when they are not numeric. The second involves the algorithms based on Collaborative Filtering [30] such as Slope one [32]. This system recommends items according to the previous ratings provided by the users [21][22][23][24][25][26][27][28].

3 Proposed System

The developed system provides the user with Information about a conference including information about the events, a schedule and a brief description as we can see in the figure 1.The system also provides information about current and upcoming events. Moreover, it provides information about the speakers, the location of the rooms, contact information of the organizers, and a section with notices and news. The system provides relevant information about nearby hotels and location of the conference. The user can calculate the route from any location to one of these sites using the GPS in the mobile device. These are the main functionalities of the system. We will now introduce the intelligent behavior of the system.

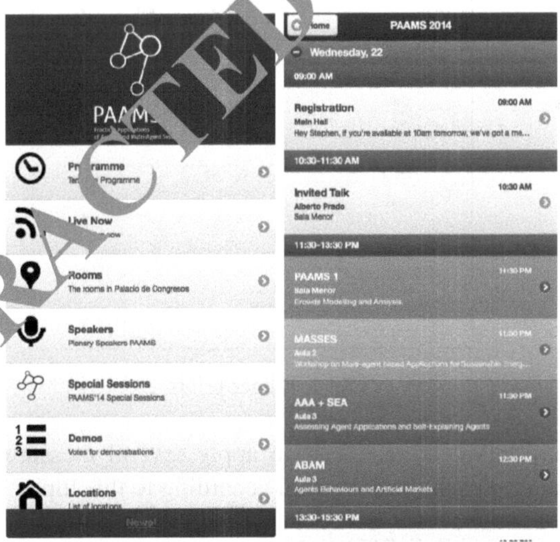

Fig. 1. a)Home screen b)Events in the conference

When the users execute the application, the system locates the user in the conference area according to the WiFi signal scanned. The WiFi signals are sent to the server, which calculates the final position of the users. The server stores the position and time for each measurement; it calculates the route of the user in the conference, the events the user has attended, and the time that spent in each one. During each session, various authors present a paper. For each paper, the author has had to include several keywords to describe the work. Using the information of the location and date of the session, the system can identify the keywords of the papers for the session the user has attended. This data can then be used to recommend additional events, sessions or papers to the user by taking the user's preferences into account, without further need to introduce additional data into the system. The process to carry out the location and recommendation is explained in the following sections.

3.1 Location System

The system uses an indoor location system based on fingerprints to locate the users and determine which events the user has attended. The first action is to create fingerprints in the facility. The facility must contain enough WiFi access points to provide a signal to each room, thus ensuring the entire facility is mapped. Figure 2 shows the fingerprints taken in the facility where the test was carried out.

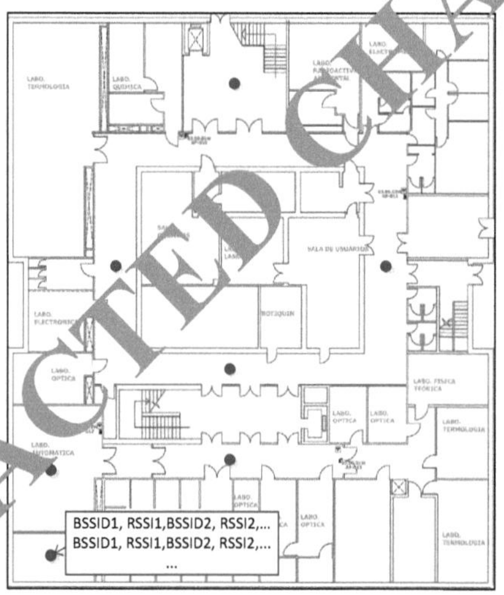

Fig. 2. Finger prints

For each fingerprints it is necessary to apply several measurements. The system obtains a set of data for each measurement according to this tuple:

$$(x, y, bssid_1, rssi_1, ..., bssid_n, rssi_n)$$

While is not necessary to obtain a high precision reading, the location region of the user should be calculated. We can therefore apply a simple algorithm such as IBK [31]. To apply the algorithm, the mobile gathers the WiFi signals. The mobile sends the BSSID and RSSI levels to the server, and the server calculates the position according to the IBK algorithm [29][30][31][32][33][34].

3.2 Collaborative Filtering with Slope One

There are many algorithms that apply distance measures with elements and determine the similarity among elements. Some of these distance measures are Euclidean, cosine, Pearson correlation etc. The main problem is that the calculation increases quadratically with the numbers of elements. The slope one [32] and the algorithm are then selected [33]. The information used during the recommendation process is shown in table 1

Table 1. Information managed by the recommender

	Key 1	…	Key m
User 1			
…			
User n			

Where v_{ij} is the value of the user i and keyword j. When the user is attending a session in a room where there is a paper being presented with the keyword k, the value of the v_{ij} is increased. In this way, the table applies the information as part of the user's preferences. The problem is that there can be keywords with similar meaning; for example, grid computing or distributed system. In this case, it is necessary to determine the user's level of interest for each keyword if some of the them are unknown. The predicted value is called p_{ij} and it is obtained using slope one according to the equations (1) and (2) . First the weighted average as the number of predictions for each keyword is calculated.

$$\overline{d}_{ij} = \frac{\sum_{k=1}^{m-1} (v_{ki} - v_{kj})}{m-1} \tag{1}$$

Where, I and j represent keywords I and j. v_{kj} represents the value of the user k with the item i, and m represents the numbers of users with known values for item i and j.

Finally the value of p_{ij} is

$$p_{ik} = \frac{\sum_{j=1}^{n-1} m_j \overline{d}_{kj}}{\sum_{j=1}^{n-1} m_j} \tag{2}$$

Where p_{ik} represents the individual i for whom the value of the keyword k is calculated. m_j is the number of known values for the keywords, and n is the number of individuals [34].

Once the values of v_{ij} and p_{ij} are known, the recommendation process is carried out. The keywords for each paper are extracted. If the value v of the keywords is unknown, then the value of p is calculated. Finally the average of these values represents the relevance of the papers to the user. Using the same process the system can calculate the relevance of a session for a user [35][36][37].

4 Results and Conclusions

The system was developed on Android devices because iOs does not allow scanning WiFi, which makes it impossible to provide the user with any type of recommendation. The system is in the testing phase and the recommendation process is not yet available in the Google play store. The system was tested in the physics building of the Faculty of Sciences at the University of Salamanca. The location was tested in the ground floor, which measures approximately 1700m^2, and contains 4 access points. These access point are from the University itself; we did not introduce new access points. By establishing the location of the user within the facility during the conference, we are able to calculate which sessions the user has attended according to the conference schedule.

To validate the system, we developed several tests within the Faculty of Sciences. We provided 39 persons with the program for the various conferences organized by the BISITE research group in Salamanca during 2013. We obtained the keywords for each paper. Next, each user selected the sessions to attend, after which the system extracted the keywords for the papers presented in that session. Table 2 stores the information about the number of times that each keyword is repeated. We removed the value of some cells in order to determine the same value using other information.

Table 2. Data managed by the recommender

	virtual organization	distributed Systems	multi-agent system	...	image processing	context aware
User 1	unknown	10	8	...	0	2
User 2	7	8	7	...	1	1
User 3	5	13	7	...	1	3
User 4	7	12	7	...	1	3
User 5	7	10	8	...	2	0
...	...	...	...		...	...
User 24	unknown	1	1	...	4	12
	...	...	...		...	...

The system is able to calculate the value of the unknown value with collaborative filtering. The predicted value for user 1 and the keyword *virtual-organization* was 6.03, while the real value was 7. Similarly, the predicted value for user 22 and the keyword *virtual organization* was 2.75, while the real value was 3. To make a better analysis we removed 5 values for each user and compared the predicted and real values; the average difference was 1.17.

The recommendation of items or sessions is carried out by analyzing the keyword of each paper. The unknown values in table 2 are calculated using slope one. After calculating the value for each of the keywords, the average is determined. This is now the overall preference value of the paper presented. To validate the performance of the system, we proceeded to delete the assistance data to a session for each user, and predict the session of interest in that time slot. In the strip, 6 parallel sessions were held at the general conference, workshops, or special sessions. The number of matches with the original decision was 82.05% but this value is highly dependent on the type of parallel sessions and user profile.

In conclusion, the system presented in this paper provides a tool that makes recommendations to users according to their own particular preferences. This system is transparent to the user and there is no need to enter additional information to calculate the recommendations. Future work will require integrating the system with easychair, creating a format to process the program, and automating the processes of extracting the information, thus allowing us to test the system during a live rather than a simulated event.

References

1. Álvarez, A., Alonso, J.M., Trivino, G.: Human activity recognition in indoor environments by means of fusing information extracted from intensity of WiFi signal and accelerations. Information Sciences 233, 162–182 (2013)
2. Fisher, J., Monahan, T.: Evaluation of real-time location systems in their hospital contexts. International Journal of Medical Informatics 81(10), 705–712 (2012)
3. Montaser, A., Moselhi, O.: RFID indoor location identification for construction projects contexts. Automation in Construction 39, 167–179 (2014)
4. Khoury, M., Kamat, R.: Evaluation of position tracking technologies for user localization in indoor construction environments. Automation in Construction 18(4), 444–457 (2009)
5. Ling, W., Li-ya, H.: Improvement of location methods based on RFID. The Journal of China Universities of Posts and Telecommunications 20(6), 36–41 (2013)
6. Galván-Tejada, C., Carrasco-Jiménez, J., Brena, R.: Bluetooth-WiFi based Combined Positioning Algorithm, Implementation and Experimental Evalua-tion. Procedia Technology 7, 37–45 (2013)
7. Farahani, S.: Chapter 3 – ZigBee and IEEE 802.15.4 Protocol Layers. ZigBee Wireless Networks and Transceivers, pp. 33–135 (2008)
8. Quiroga, L., Mostafa, J.: An experiment in building profiles in information filtering: the role of context of user relevance feedback. Information Processing & Management 38(5), 671–694 (2002)
9. Corchado, J.M., Fyfe, C.: Unsupervised neural method for temperature forecasting. Artificial Intelligence in Engineering 13(4), 351–357 (1999)
10. Fdez-Riverola, F., Corchado, J.M.: CBR based system for forecasting red tides. Knowledge-Based Systems 16(5), 321–328 (2003)
11. Tapia, D.I., Abraham, A., Corchado, J.M., Alonso, R.S.: Agents and ambient intelligence: case studies. Journal of Ambient Intelligence and Humanized Computing 1(2), 85–93 (2010)

12. Corchado, J.M., Lees, B.: Adaptation of cases for case based forecasting with neural network support. In: Soft Computing in Case Based Reasoning, pp. 293–319 (2001)
13. Corchado Rodríguez, J.M.: Redes Neuronales Artificiales: un enfoque práctico. Servicio de Publicacións da Universidade de Vigo, Vigo (2000)
14. Bajo, J., Corchado, J.M.: Evaluation and monitoring of the air-sea interaction using a CBR-agents approach. In: Muñoz-Ávila, H., Ricci, F. (eds.) ICCBR 2005. LNCS (LNAI), vol. 3620, pp. 50–62. Springer, Heidelberg (2005)
15. Fraile, J.A., Bajo, J., Corchado, J.M., Abraham, A.: Applying wearable solutions in dependent environments. IEEE Transactions on Information Technology in Biomedicine 14(6), 1459–1467 (2011)
16. Corchado, J.M., De Paz, J.F., Rodríguez, S., Bajo, J.: Model of experts for decision support in the diagnosis of leukemia patients. Artificial Intelligence in Medicine 46(3), 179–200 (2009)
17. De Paz, J.F., Rodríguez, S., Bajo, J., Corchado, J.M.: Case-based reasoning as a decision support system for cancer diagnosis: A case study. International Journal of Hybrid Intelligent Systems 6(2), 97–110 (2009)
18. Tapia, D.I., Rodríguez, S., Bajo, J., Corchado, J.M.: FUSION@, a SOA-based multi-agent architecture. In: International Symposium on Distributed Computing and Artificial Intelligence (DCAI 2008), pp. 99–107 (2008)
19. Corchado, J.M., Aiken, J.: Hybrid artificial intelligence methods in oceanographic forecast models. IEEE Transactions on Systems, Man and Cybernetics, Part C: Applications and Reviews 32(4), 307–313 (2002)
20. Corchado, J.M., Aiken, J., Rees, N.: Artificial intelligence models for oceano-graphic forecasting. Plymouth Marine Laboratory (2001)
21. Rodríguez, S., Pérez-Lancho, B., De Paz, J.F., Bajo, J., Corchado, J.M.: Ovamah: Multiagent-based adaptive virtual organizations. In: 12th International Conference on Information Fusion, FUSION 2009, pp. 990–997 (2009)
22. Tapia, D.I., De Paz, J.F., Rodríguez, S., Bajo, J., Corchado, J.M.: Multi-agent system for security control on industrial environments. International Transactions on System Science and Applications Journal 4(3), 222–226 (2008)
23. Borrajo, M.L., Baruque, B., Corchado, E., Bajo, J., Corchado, J.M.: Hybrid neural intelligent system to predict business failure in small-to-medium-size enter-prises. International Journal of Neural Systems 21(04), 277–296 (2011)
24. De Paz, J.F., Rodríguez, S., Bajo, J., Corchado, J.M.: Mathematical model for dy-namic case-based planning. International Journal of Computer Mathematics 86(10-11), 1719–1730 (2009)
25. Bajo, J., De Paz, J.F., Rodríguez, S., González, A.: Multi-agent system to monitor oceanic environments. Integrated Computer-Aided Engineering 17(2), 131–144 (2010)
26. Tapia, D.I., Alonso, R.S., De Paz, J.F., Corchado, J.M.: Introducing a distributed architecture for heterogeneous wireless sensor networks. In: Omatu, S., Rocha, M.P., Bravo, J., Fernández, F., Corchado, E., Bustillo, A., Corchado, J.M. (eds.) IWANN 2009, Part II. LNCS, vol. 5518, pp. 116–123. Springer, Heidelberg (2009)
27. Rodríguez, S., de Paz, Y., Bajo, J., Corchado, J.M.: Social-based planning model for multiagent systems. Expert Systems with Applications 38(10), 13005–13023 (2011)
28. Pinzón, C.I., Bajo, J., De Paz, J.F., Corchado, J.M.: S-MAS: An adaptive hierarchical distributed multi-agent architecture for blocking malicious SOAP messages within Web Services environments. Expert Systems with Applications 38(5), 5486–5499
29. Corchado, J.M., Bajo, J., De Paz, J.F., Rodríguez, S.: An execution time neural-CBR guidance assistant. Neurocomputing 72(13), 2743–2753 (2009)

30. Zhang, J., Peng, Q., Sun, S., Liu, C.: Collaborative filtering recommendation algorithm based on user preference derived from item domain features. Physica A: Statistical Mechanics and its Applications 396, 66–76 (2014)
31. Aha, W., Kibler, K., Albert, M.: Instance-based learning algorithms. Machine Learning 6, 37–66 (1991)
32. Lemire, D., Maclachlan, A.: Slope One predictors for online rating-based col-laborative filtering. In: Proceedings of SIAM Data Mining, SDM 2005 (2005)
33. Campillo-Sánchez, P., Botía, J., Gómez-Sanz, J.: Development of Sensor Based Applications for the Android Platform: an Approach Based on Rea-listic Simulation. Advances in Distributed Computing and Artificial Intelligence Journal 1(4), 23–28 (2013)
34. Wang, X., Tarrío, P., Metola, E., Casar, J.: User-independent accelerometer-based gesture recognition for mobile devices. Advances in Distributed Computing and Artificial Intelligence Journal 1(3), 11–25 (2012)
35. Sánchez-Pi, N., Carbó, J., Molina, J.M.: A Knowledge-Based System Approach for a Context-Aware System. Knowledge-Based Systems (27), 1–17 (201[illegible])
36. Gómez, J., Patricio, M.A., García, J., Molina, J.M.: Communication in distributed tracking systems: an ontology-based approach to improve cooperation. Expert Systems 28(4)
37. Serrano, E., Gómez-Sanz, J.J., Botía, J.A., Pavón, J.: Intelligent data analysis applied to debug complex software systems. Neurocomputing 72(13), 2[illegible]–2795 (2009)

A Multi-agent System for Infrared and Color Video Fusion

José Manuel Gascueña[1], Juan Serrano-Cuerda[1], José Carlos Castillo[2], Antonio Fernández-Caballero[1], and María T. López[1]

[1] Universidad de Castilla-La Mancha, Departamento de Sistemas Informáticos & Instituto de Investigación en Informática de Albacete, 02071-Albacete, Spain
Antonio.Fdez@uclm.es

[2] Universidad Carlos III de Madrid, Departamento de Ingeniería de Sistemas y Automática, Av. Universidad 30, 28911-Leganés, Spain

Abstract. Nowadays, the use of heterogeneous cameras is preferred to develop robust monitoring systems. Under multi-camera setups, information fusion processes are required to improve the results independently obtained from each installed camera. In addition, multi-agent systems are designed to face up the development of complex systems such as advanced monitoring systems. This paper describes the use of INT^3 technologies to develop an agent-based robust people detection and tracking system completed by a scheme for infrared and color video fusion. The system development process covers from the modeling to the implementation of the system. The meaning of the levels selected by an analyst is detailed as well as the components required to fulfil the requirements for human detection and tracking. The proposed approach has the ability to make decisions about the convenience to fuse visible and infrared spectra information.

Keywords: Multi-agent system, Color video, Infrared video, Video fusion, INT^3 technologies.

1 Introduction

Information obtained from a monitoring system composed of a single type of camera is usually not enough to detect and track humans under changeable environment conditions [16]. It is due to problems arising from changes in illumination [17] and/or temperature [10] occurring in the environment. Therefore, the obvious need for different spectra information fusion appears to allow the development of systems capable of performing monitoring tasks in a more efficient way.

Furthermore, the management and processing of information obtained from monitoring systems is a very complex problem. These systems are usually composed of a series of heterogeneous devices and distributed along the monitored environment. The incorporation of software components, agents, with a certain degree of autonomy, that also are able to collect information from the environment and to cooperate to carry out object detection and tracking tasks, is proposed a solution to face up the inherent complexity of information fusion systems [2]. The characteristics of autonomy and cooperation are often cited as the reason of why multi-agent systems are specially suitable to carry out detection and tracking tasks [8], [14], [13].

J.B. Perez et al. (eds.), *Trends in Practical Applications of Heterogeneous Multi-agent Systems. The PAAMS Collection*, Advances in Intelligent Systems and Computing 293,
DOI: 10.1007/978-3-319-07476-4_16,

This paper describes the development of a multi-agent monitoring system using INT^3 technologies [9], [7], which merges agent and sensor fusion technologies to perform a robust human detection, employing information from color and infrared cameras. The rest of the paper is structured as follows. Section 2 provides an overview of INT^3 framework with its main features. Section 3 offers the development details of the multi agent system for information fusion. Finally, the most relevant conclusions are offered.

2 INT^3 Technologies

INT^3 is the acronym to refer a set of technologies applied for the development of applications capable of carrying out a multisensory monitoring and *INT*erpretation of behaviors and situations for an *INT*elligent *INT*ervention in complex and dynamic environments. These technologies are (i) a domain specific modeling language to model this kind of applications (INT^3*-DSML*) [9], (ii) a software development process to develop them (INT^3*-SDP*) [9], and (iii) a framework to implement them (INT^3*-Horus*) [7]. Currently, applications developed putting into practice these technologies possess some common features which will be described next.

2.1 Main Features of INT^3-DSML

The domain specific modeling language, INT^3*-DSML*, is able to deal with the challenges of state-of-the-art monitoring systems providing the applications with a series of distinguishing features:

- **Multisensory**. The language supports modeling applications with different sources of information (e.g. color cameras, infrared cameras, volumetric sensors, contact sensors, wireless sensor networks).
- **Multilevel**. A level is a processing layer which includes a set of software components representing specific algorithms. They are responsible for carrying out the functionality associated to the level by the developer. For each level, there is a set of inputs and outputs, which should be fulfilled by the software components. Nevertheless, a software component may not cover all inputs and outputs associated to its level but the subset that best fits its operation and requirements. However, the global collection of components related to each level is constrained to the inputs and outputs belonging to that level, though the architecture is flexible in the sense that the set of inputs and outputs can be extended if application requirements demands it.
- **Information Fusion Capability**. Some levels needed are devoted to perform information fusion tasks from different components. Thus, the language also supports the modeling of the concept of data fusion. The main advantage of putting into practice this concept is that data quality is usually improved as a result of information fusion processes. Approaches without fusion techniques lack of these enhancements.

2.2 Main Features of *INT*^{3}*-SDP*

INT^3*-SDP* is the process to develop systems for multisensory monitoring and activity interpretation. Let us remark that models created along the process use the language *INT*^{3}*-DSML*. *INT*^{3}*-SDP* consists of two phases:

- *Modeling the Monitoring Domain.* The main goal of this phase is to describe the customers' necessities for multisensory monitoring.
- *Modeling the Monitoring Deployment.* This phase focuses on the specification and implementation of software components which perform the monitoring and activity interpretation tasks.

2.3 Main Features of the *INT*^{3}*-Horus* Framework

INT^3*-Horus* provide the developers not only with several tools to easier the integration of code into the framework but also with a set of operation modules to generate flexible monitoring applications. Some of the main aspects are detailed next:

- **Model-View-Controller**. The implementation of monitoring applications in the framework is based on the approach proposed in software pattern Model-View-Controller (MVC) [15]. Several design decisions were taken by the architects of *INT*^{3}*-Horus* to provide flexibility at the time to incorporate new components into the framework without involving a major change in the architecture. On the one hand, the *Controller*, playing the conductor role, invokes the functionalities of new execution blocks proposed by the framework architects, which are called "algorithms", regarding the original approach suggested for the MVC pattern. Examples of these new blocks are components related to business logic, communications and database access, as well as traditional images processing algorithms (e.g. computer vision algorithms to perform tracking, image acquisition, segmentation ...). On the other hand, each component has a local model responsible for storing its local data and providing primitives to manage them. Finally, a common model has a global view of the application data, that is, it stores information generated or required by the levels. So, the common model is another novel concept regarding the original MVC since it implements both the controller and the global data model.
- **Multilevel**. The *INT*^{3}*-Horus* framework is also implemented according to concepts which constitute the modeling language *INT*^{3}*-DSML*, thus providing traceability between modeling and implementation. The framework levels establish a hierarchy from the lower level (information acquisition) to the upper ones. In this sense, the framework advantage of layer pattern "enhancing maintainability, extensibility, and reusability of the system" [18]. These advantages complement the three benefits of using MVC pattern [4]: (1) separation of concerns in the code base, (2) developer specialization and focus, and (3) parallel development by separate teams.
- **Hybrid**. Monitoring applications implemented using the *INT*^{3}*-Horus* framework are categorized as a hybrid distributed systems. Remote nodes are responsible for information processing related to lower levels, whilst a central node is responsible for collecting information from remote nodes, fusing it and to perform high level

operation. It also houses a global controller responsible for several tasks: (1) initializing the model and view, (2) loading and connecting components, (3) managing the processing stack of the components, (4) monitoring components' behavior, and, (5) swapping a component in runtime.
- **Multisensory**. *INT*^{3}*-Horus* framework provides support to work with different types of physical devices (e.g. video encoders, webcams, wireless sensors networks, and so on) and guidelines and support to incorporate new ones.

INT^3 technologies have already been applied to the development of applications involving topics related to ambient intelligence area using agents and data fusion. For example, *INT*^{3}*-Horus* was selected to implement systems for detecting stress [1] and elderly care [3]. Moreover *INT*^{3}*-SDP* was applied to develop a fall detection system [9]. In these applications and according to their behavior, components belonging to levels hierarchy (modules), as well as the global controller, can be considered as reactive agents [19].

3 Development of the Human Detection and Tracking System

The first phase of *INT3-SDP*, *Modeling the Monitoring Domain*, consists in describing the environment and the sensors along with their location and associated budget. Two cameras are selected by the analyst, a person with a high degree of knowledge about monitoring systems technologies and *INT3-DSML* terminology, for the development of a robust people detection system. On the one hand, a *FLIR A-320* infrared camera was configured as follows: (1) dynamic temperature range, and (2) image acquisition at a resolution of 320×240 pixels with a frame rate of 5 frames per second (fps). On the other hand, a *SONY FCB-EX780bp* color camera was set up to acquire frames at a resolution of 320×240 pixels with a frame rate of 5 fps. This frame rate is forced to synchronize the acquired frames from both cameras. Cameras are placed in parallel and focused to a common point of the same outdoor scenario in order to obtain two similar views of the same scene. This information is used to create a *Sensor Model*, a *Physical Sensor Model*, a *Budget Model* and a *Environment Model*.

Once the domain model has been established, the *Modeling the Monitoring Deployment* phase begins. In this case, the analyst identified the conceptual levels required to model the system as well as software components required to carry out the processing of each level. These requirements are summarized as follows:

- *Acquisition level* simultaneously acquires (and synchronizes) video grabbed from an infrared camera and a visible camera. The two components to model this functionalities are *IRImageAcquisition* and *ColorImageAcquisition*, respectively.
- *Segmentation* level detects human candidates in infrared and visible spectra. To achieve this goal, a series of segmentation algorithms on the infrared and visible acquired images are required. Specifically, *IRImageSegmentation* and *ColorImage-Segmentation* components model approaches of segmentation to segment images provided by *IRImageAcquisition* and *ColorImageAcquisition* components, respectively. Specifically, segmentation algorithms chosen by the expert are described in [5] and [6], respectively.

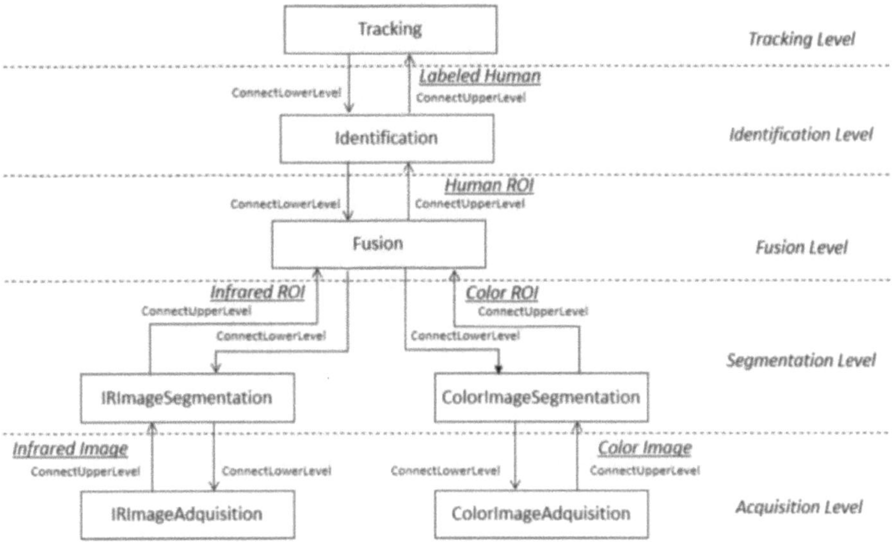

Fig. 1. Configuration Model

- *Fusion* level carries out the fusion of those human candidates which were segmented by the previous level. The results obtained in this level are regions of interest (ROIs) associated to the human candidates found in the segmentation level. Section 3.1 describes the analyst's fusion proposal.
- *Identification level* is responsible for generating a list of the humans detected along the image sequence with their associated features (location, height, width).
- *Tracking level* calculates humans trajectories and speed based on their last locations.

The analyst is also responsible for other relevant tasks in this second phase: identifying the number of component instances required for the final application, as well as their connections. These tasks contribute to the generation of a *Configuration Model* (see Fig. 1). On the other hand, the analyst uses information about the levels, inputs, outputs, and the necessary components to create a *Component Repository Model.*

3.1 Information Fusion Level Description

An approach based on *region level fusion* [12] was chosen by the analyst to fuse segmented images. These approaches are an interesting option when working with different image spectra [11]. One of the advantages is the possibility of using intelligent fusion rules regarding the specifics features of each spectrum. Thus, sharpness problems are mitigated in low lightning and warm environmental conditions thanks to the use of complementary information from both cameras combined with intelligent fusion rules. The fusion algorithm is described taking into account those situations where information fusion is not suitable. In these cases, the ROIs obtained by the components located at

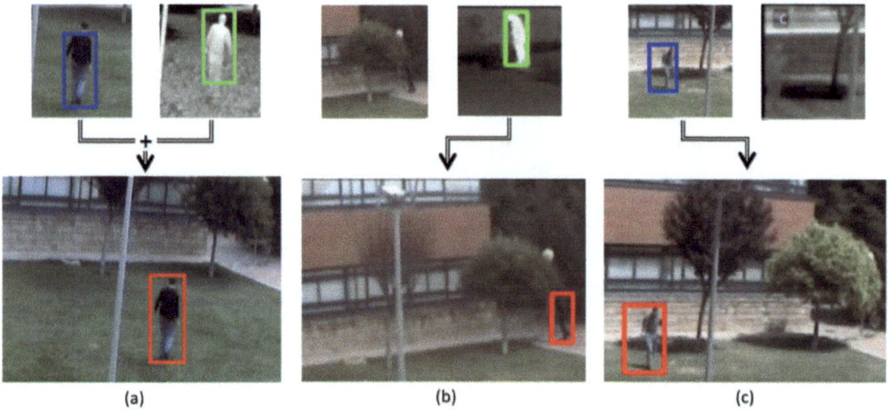

Fig. 2. Human detection using (a) infrared and color fusion, (b) infrared, (c) color

segmentation level (*IRImageSegmentation* and *ColorImageSegmentation*) are the information provided to the *Identification* component.

Once the calibration has been performed, the fusion process begins. Apart from the human candidates given by visible and infrared segmentation, those results only found by one of the approaches are also used to obtain ROIs related to humans. Each ROI is accepted depending on image properties such as the mean illumination and the standard image deviation for the infrared spectrum, whilst the average intensity is the main cue in the visible one. The analyst made two hypothesis in order to decide whether or not fusion must be applied to obtain a ROI:

- Sometimes the infrared spectrum not only works better than the visible information on dark situations but also on zones covered by a great amount of shades. In this situations, if the infrared image properties provide enough confidence, which is calculated for the *IRImageAcquisition* component, the *Identification* component consumes the ROI obtained in the *IRImageSegmentation* one.
- Sometimes humans found in the infrared scene are very difficult to distinguish from the background. Yet, humans can be easily detected in the visible spectrum. In this case, the properties of the visible spectrum result in the attribution of a high confidence value to this segmentation. Thus, ROIs obtained by *ColorImageSegmentation* are the input of the *Identification* component.

In both previous particular cases, it can be said that *Fusion* component decides do not carry out the segmented images fusion. In the cases in which *Fusion* component decides do not carry out the segmented images fusion, the ROI obtained at the segmentation level producing better results (in infrared or visible spectra) feeds the *Identification* component, according to the casuistry previously described.

Finally, once the system has been modeled, *INT*^{3}*-Horus* framework served as a basis for implementation. Fig. 2 shows different results of a human detected in different locations of the environment. They are related to the three possibilities described previously, that is, applying fusion (Fig. 2a), using only the results of infrared segmentation (Fig. 2b) and using only the results of color segmentation (Fig. 2c), respectively.

4 Conclusions

This paper illustrates an application of INT^3 technologies to develop a specific agent-based human detection and tracking system which uses information fusion of visible and infrared spectra video. These technologies are a domain specific modeling language, INT^3*-DSML*, a software development process, INT^3*-SDP*, and an implementation framework, INT^3*-Horus*.

The system development process covers from the modeling to the implementation of the system. The meaning of the levels selected by an analyst is detailed as well as the components required to fulfil the requirements of human detection and tracking applications. The analyst is an expert person with extensive knowledge and skills about particular monitoring area. Therefore, it can be affirmed that INT^3 technologies allow to model and implement monitoring applications. These applications include reactive agents to coordinate the processing of information required to carry out monitoring and activity interpretation tasks according to the approaches proposed by experts in the field.

Acknowledgment. This work was partially supported by Spanish Ministerio de Economía y Competitividad / FEDER under TIN2010-20845-C03-01 grant.

References

1. Carneiro, D., Castillo, J.C., Novais, P., Fernández-Caballero, A., Neves, J.: Multimodal behavioral analysis for non-invasive stress detection. Expert Systems with Applications 39(18), 13376–13389 (2012)
2. Castanedo, F., García, J., Patricio, M.A., Molina, J.: A multi-agent architecture based on the BDI model for data fusion in visual sensor networks. Journal of Intelligent and Robotic Systems 62(3-4), 299–328 (2011)
3. Costa, A., Castillo, J.C., Novais, P., Fernández-Caballero, A., Simoes, R.: Sensor-driven agenda for intelligent home care of the elderly. Expert Systems with Applications 39(15), 12192–12204 (2012)
4. Chiaretta, S., Nayyeri, K.: Beginning ASP.NET MVC 1.0. Wiley Publishing, Inc. (2009)
5. Fernández-Caballero, A., Castillo, J.C., Serrano-Cuerda, J., Maldonado-Bascón, S.: Real-time human segmentation in infrared videos. Expert Systems with Applications 38(3), 2577–2584 (2011)
6. Fernández-Caballero, A., López, M.T., Carmona, E.J., Delgado, A.E.: A historical perspective of algorithmic lateral inhibition and accumulative computation in computer vision. Neurocomputing 74(8), 1175–1181 (2011)
7. Fernández-Caballero, A., Castillo, J.C., López, M., Sokolova, M.V.: INT^3-Horus framework for multispectrum activity interpretation in intelligent environments. Expert Systems with Applications 40(17), 6715–6727 (2013)
8. Gascueña, J.M., Fernández-Caballero, A.: Agent-oriented modeling and development of a person-following mobile robot. Expert Systems with Applications 38(4), 4280–4290 (2011)

9. Gascueña, J.M., Castillo, J.C., Navarro, E., Fernández-Caballero, A.: Engineering the development of systems for multisensory monitoring and activity interpretation. Int. J. of Systems Science 45(4), 728–740 (2014)
10. Hammoud, R.I.: Augmented vision perception in infrared: Algorithms and applied systems. Springer (2009)
11. Lewis, J.J., O' Callaghan, R.J., Nikolov, S.G., Bull, D.R., Canagarajah, N.: Pixel- and region-based image fusion with complex wavelets. Information Fusion 8(2), 119–130 (2007)
12. Luo, R.C., Kay, M.G.: Data fusion and sensor integration: State of the art 1990s. In: Abidi, M.A., Gonzalez, R.C. (eds.) Data Fusion in Robotics and Machine Intelligence, pp. 7–135. Academic Press, San Diego (1992)
13. Muñoz-Salinas, R., Aguirre, E., García-silvente, M., Ayesh, A., Góngora, M.: Multi-agent system for people detection and tracking using stereo vision in mobile robots. Robotica 27(5), 715–727 (2009)
14. Pavón, J., Gómez-Sanz, J., Fernández-Caballero, A., Valencia-Jiménez, J.J.: Development of intelligent multisensor surveillance systems with agents. Robotics and Autonomous Systems 55(12), 892–903 (2007)
15. Reenskaug, T.: Thing-Model-View-Editor - An example from a planning system. Technical note, Xerox PARC (1979)
16. Serrano-Cuerda, J., María, T., López, M., Fernández-Caballero, A.: Robust Human Detection and Tracking in Intelligent Environments by Information Fusion of Color and Infrared Video. In: 7th Int. Conf. on Intelligent Environments, pp. 354–357 (2011)
17. Stefano, L., Tombari, F., Lisi, E., Mattoccia, S.: Robust and accurate change detection under sudden illumination variations. In: Workshop on Multi-dimensional and Multi-view Image Processing, pp. 103–109 (2007)
18. Weyns, D.: Architecture-based design of multi-agent systems. Springer (2010)
19. Wooldridge, M.: An Introduction to MultiAgent Systems, 2nd edn. John Wiley & Sons (2009)

The Impact of Recommendation Agents' Type of Voice on Perceived Social Presence, Trust and Users Intentions on an Insurance Website

Emna Cherif[1] and Jean-François Lemoine[1,2]

[1] Université Paris 1 Panthéon – Sorbonne, Paris, France
Cherif_emna@yahoo.fr
[2] ESSCA School of management, Angers, France
jflemoine30@hotmail.com

Abstract. This paper aims to demonstrate the influence that recommendation agent type voice may have on users reactions. Through experimentation, we compare the effects of human and synthetic voice on perceived social presence, recommendation agent trust, website trust and behavioral intentions. The findings suggest that the human voice is likely to provide a higher level of social presence and recommendation agent trust. The structural equation model shows that social presence has a positive effect both on recommendation agent trust and behavioral intentions. In turn, recommendation agent trust influences website trust and behavioral intentions. Finally, results show that website trust positively impacts behavioral intentions.

Keywords: recommendation agent, human vs. synthetic voice, social presence, trust.

1 Introduction

Recommendation agents are becoming increasingly present in websites. These features are designed to help and guide users during their online experience. Recommendation agents[1] are defined "as general graphic representations that are personified by means of computer technology" (Holzwarth, Janiszewski, & Neumann, 2006).

Most research on recommendation agents has demonstrated that their presence makes for a more intimate experience, humanizes the interaction between the user and the site, simulates a social presence on the recommendation interfaces, and enhances trust (Holzwarth, Janiszewski, & Neumann, 2006 ; Keeling, McGoldrick, & Beatty, 2010 ; Kohler, Rohm, Ruyter, & Wetzels, 2011 ; Qiu & Benbasat, 2009b ; Wang, Baker, Wagner, and Wakefield, 2007).

However, aside from the positive impact a recommendation agent's presence may have on the interaction, the most important aspect is the way in which recommenda-

---

[1] Multiple terms are used: avatars (Holzwarth, Janiszewski, & Neumann, 2006), conversational embodied agent (Cassell & Bickmore, 2000) ,virtual agent, interactive character, shopping agent and animated agent, etc. (McGoldrick, Keeling, & Beatty, 2008).

J.B. Perez et al. (eds.), *Trends in Practical Applications of Heterogeneous Multi-agent Systems. The PAAMS Collection*, Advances in Intelligent Systems and Computing 293,
DOI: 10.1007/978-3-319-07476-4_17, © Springer International Publishing Switzerland 2014

tion agent's they are designed to interact with the user. This addresses the issue of recommendation agent anthropomorphism. Anthropomorphism is "the tendency to attribute human characteristics to inanimate objects, animals and others with a view to helping us rationalize their actions" (Duffy, 2003). Burgoon et al. (2000) says that agents can be designed in a more or less anthropomorphic way, giving them human characteristics such as voice, facial expressions and human gestures. Indeed, a recent qualitative study demonstrated the importance of anthropomorphic characteristics of recommendation agents in relation to user perceptions and reactions (Lemoine & Cherif, 2012). One of the most cited characteristics was the importance of the voice style. Moreover, research on social response theory showed users react to a computer with different voices as if they were talking with different people. Similarly, when speaking with different computers using the same voice, customers reacted as if they were still speaking with the same person (Steuer & Nass, 1993).

Thus, this study aims to understand how the type of voice (human versus synthetic) of the recommendation agent could influence the perceived social presence, recommendation agent trust, website trust and behavioral intentions. While numerous studies have examined the effects that anthropomorphism may have on a recommendation agent, little attention has been given to how an agent's voice type may affect a potential customer's behavior towards an agent.

In regard to its practical implications, this study tries to provide conception guidance to recommendation agent designers that may engage positive behaviors.

First, this study focuses on the importance of recommendation agents' design. It then discusses perceived social presence, trust and intentions, and considers how these factors could be influenced by the voice type (human versus synthetic) of the recommendation agent. After presenting the research design, the analysis is presented. Finally, results are then presented and discussed.

2 Literature Review

With the advent of the internet, managers and researchers have begun to pay special attention to the design of the websites atmosphere. In order to provide users with an enjoyable online experience, it is necessary to improve the websites atmosphere by including social factors. Thus, to address the lack of human warmth and sociability often felt when interacting in online environments (Holzwarth, Janiszewski, & Neuman, 2006), managers and site designers are increasingly trying to provide web interfaces with different attributes that enhance the social presence on a site, such as the addition of music, chat and recommendation agents (Wang, Baker, Wagner, & Wakefield, 2007 ; Holzwarth, Janiszewski, & Neumann, 2006 ; Lee, Jung, Kim, & Kim, 2006).

Indeed, research has shown that the presence of a recommendation agent simulates a social presence on the internet, humanizes the interaction between the user and the site, and makes the interaction more personal (Holzwarth, Janiszewski, & Neumann, 2006 ; Kohler, Rohm, Ruyter, & Wetzels, 2011 ; Qiu & Benbasat, 2009b ; Wang, Baker, Wagner, & Wakefield, 2007). Moreover, it has been shown that recommendation agent enhances trust (Keeling, McGoldrick, & Beatty, 2010) and perceived enjoyment and pleasure (Qiu & Benbasat, 2009b ; Wang, Baker, Wagner, & Wakefield, 2007). Furthermore, researchers have continued studying the effect of a recommendation agent's anthropomorphism, like gesture, voice, facial expressions, on users reactions (Cowell & Stanney, 2005 ; Gulz & Haake, 2006 ; Nowak & Rauh, 2008).

In this study, we focus specifically on the effect of the voice type : human versus synthetic voice of a recommendation agent on an insurance website providing recommendations. Synthetic voice is a technology that, among its many uses, allows for more effective communication among speech impaired individuals (Stern, Mullennix, & Yaroslavsky, 2006). Research mainly focuses on the voice effects on impaired users (Loiacono et al., 2013; Grichkovtsova et al., 2012; Wongkia et al., 2012). D'Alessandro (2001, 2006) shows that text-to-speech voice should not only be intelligible but also expressive and emotional. Therefore, researchers and developers have spent countless hours in an effort to optimize the voice and prosodic qualities.

Previous research on Human –Computer Interaction showed the importance of the recommendation agent voice on user reactions such as social presence and trust, and its ability to reinforce positive reactions in relation to the credibility, the attractiveness, the positive intentions and the perceived personality of the system (Evans & Kortum, 2010; Galanxhi & Nah, 2007; Lee & Nass, 2003; Qiu & Benbasat, 2005; Stern, Mullennix, & Yaroslavsky, 2006; Edlund, Gustafson, Heldner & Hjalmarsson, 2008).

Stern et al. (1999) showed that a human voice is perceived as more persuasive than a synthetic voice.

According to these results, we expect that, in contrast to using a synthetic voice, the human voice will create a more positive user response in terms of perceived social presence, recommendation agent trust and website trust.

H 1a: Users' evaluation of social presence on website is higher with human voice than synthetic voice.

H 1b: Users' trust in the recommendation agent is higher with human voice than synthetic voice.

H 1c: Users' trust in the website is higher with human voice than synthetic voice.

Qiu and Benbasat (2009b) show that social presence positively influences trusting beliefs and behavioral intentions. Consequently, we propose:

H 2: Social presence perception will directly and positively influences trust in recommendation agent

H 3: Positive social presence perception will lead to more positive intentions

While much research has explored the impact of recommendation agent's presence on users' trust (Keeling, McGoldrick, & Beatty, 2010; Galanxhi & Nah, 2007), only some research has focused on the users' trust in the recommendation agent (Qiu & Benbasat, 2005 ; Wang & Benbasat, 2005, 2008).

Thus, we can logically conclude that the user's website trust could depend on the presence of the recommendation agent on the website. Indeed, if trusting the website could somehow rely on the presence of the recommendation agent, one must first trust the recommendation agent. It contributes to the user website trust and engages more positive intentions.

H 4: Trust in recommendation agents will directly and positively influences trust in the website

H 5: Trust in recommendation agents will directly and positively influences behavioral intentions

H 6: Trust in the website will directly and positively influences behavioral intentions

3 Research Design: Sample , Procedure and Measures

The recommendation agent used in the experiment was a recommender system for insurance products. It was designed according to two experimental conditions of our study: the human voice and the synthetic voice[2].

Participants were recruited from a panel of internet users. They were invited to surf existing insurance website with a spoken recommendation agent (figure 1) and simulate information searching. They were randomly allocated to surf on a website with one of the two recommendation agent's voice type. After two minutes of exposition on the website to ensure that all of the respondents used the recommendation agent, they completed an online questionnaire. All the participants were asked to provide the name of the recommendation agent, while we only kept the responses of those who answered correctly. This question allowed us to verify whether or not each participant actually heard the agent. The sample consisted of 589 participants. 319 of them were exposed to the human voice version of the recommendation agent and 270 were exposed to the synthetic voice version. The average age of respondents was 42 years (SD=9,87) with equal proportions of women and men (297/292). Given the nature of the website, all of them were employed. The average income was €2500 per month. In order to avoid potential biases, the proportionality of each group was respected.

Fig. 1. Screen shot of the recommendation agent

[2] Authors thank Davi Interactive and GMF, the companies which provide them with the website and the two experimental conditions of the virtual agent.

All scales used are validated in existing literature. Social presence was measured using Qiu & Benbasat (2009a) seven point scale (1= disagree completely, 7 agree completely). Recommendation agent trust was based on Qiu & Benbasat (2009a) measure instrument and website trust was measured using Keeling, McGoldrick, & Beatty (2010) scale. The scales were also rated on a 7 point Likert scales. Finally, three items scales measuring intentions were used (Keeling, McGoldrick, & Beatty, 2010). All measurement items are presented in appendix (with minor changes in words). Reliability was good and uni-dimensionality was confirmed for each : social presence[3] (α=0,936 ; 83,865% of the variance), recommendation agent trust (α=0,955 ; 71,246% of the variance) and website trust (α=0,911 ; 74,025% of the variance).

4 Analyses and Results

Once a scales reliability and validity had been assessed, summated scales based on the factor scores of the items were used in ANOVA analyses. In order to examine the effects of a recommendation agent's voice, the significance and validity of Levene's test was further examined. Its non-significance was assumed for all the performed analysis.

To test the first three hypotheses (H1a, H1b, H1c) for the impact of voice type (human versus synthetic) on perceived social presence, recommendation agent trust and website trust, data was analyzed using several univariate ANOVAs. We performed a series of analyses of variance with the voice type as the main independent variable.

Significant effects of voice type were seen for perceived social presence ($F(1, 587) = 23{,}281$; $p = .000$). The human voice condition created the highest levels of perceived social presence ($M = 0{,}179$), followed by the synthetic voice condition ($M = -0{,}212$). Thus, H1b was supported. Voice type had a significant main effect on recommendation agent trust ($F(1, 581) = 5{,}635$; $p = .018$). The human voice generated the most positive perceived social presence ($M = 0{,}089$), followed by the synthetic voice ($M= -0{,}105$). Therefore, H1b was again supported. Voice type had no significant effect on website trust, thus, H1c was not supported.

In order to examine the effects of perceived social presence, recommendation agent trust and website trust on behavioral intentions, we used AMOS to perform a structural equation model.

Structural equation analysis was performed to test H2, H3, H4, H5, H6.

First, we examined the measurement model using a confirmatory factor analysis. Indeed, before estimating the structural model relationships, we should observe the factor loadings and convergent validity, reliability and discriminant validity of each construct. As shown in table 1, all the path coefficients from the latent variable to observed variables are statistically significant, suggesting convergent validity. The discriminant validity is assessed when the average variance extracted (AVE) for each

---

[3] The item "I felt a sense of personalness in the agent" was removed from the scale of social presence because it has an extremely low factor loading.

factor should be higher than the squared pairwise correlations between factors (Fornell and Larcker 1981) (table 2). Moreover, Kline (1998) suggests that all correlations values were between 0.1 and 0.85 in order to support discriminant validity.

Moreover, goodness of fit indices were highly acceptable ($\chi2$ (df) =3.380; GFI=0.898; CFI=0.957; TLI=0.951 and RMSEA=0.064).

Moreover, goodness of fit indices were highly acceptable ($\chi2$ (df)=3.380 , GFI=0.898, CFI=0.957 , TLI=0.951 , and RMSEA=0.064).

Table 1. Results of confirmatory factor analysis

Measures	Estimate	t-value	SMC	Reliability
SP* 1	0.906		0.821	0,980
SP*2	0.888	32.920	0.788	
SP*3	0.827	28.085	0.683	
SP*4	0.925	36.310	0.856	
WT**1	0.868		0.753	0,914
WT**2	0.758	22.479	0.574	
WT**3	0.818	25.541	0.669	
WT**4	0.914	31.365	0.834	
WT**5	0.754	22.329	0.569	
I****1	0.889		0.064	0,684
I****2	0.729	19.159	0.531	
I****3	0.253	5.858	0.790	
RA T***1	0.743		0.553	0,958
RA T***2	0.812	20.549	0.659	
RA T***3	0.856	21.831	0.733	
RA T***4	0.848	21.591	0.720	
RA T*** 5	0.842	21.417	0.710	
RA T***6	0.839	21.339	0.704	
RA T***7	0.816	20.671	0.666	
RA T***8	0.848	21.590	0.719	
RA T***9	0.809	20.484	0.655	
RAT***10	0.821	20.808	0.674	

SP*: Social presence; WT**: Website trust; RA T***: Recommendation agent trust; I****: Intentions

All the proposed relationships in the model were statistically significant (p<0, 01). Perceived social presence has positive effects on trust in recommendation agent (β=0.504, t=15.820, p=0.000) and on behavioral intentions (β=0.177, t=5.022, p=0.000). Moreover, the results of AMOS analysis indicates that trust in recommendation agent has a positive influence on website trust in one hand (β=0.690, t=18.127, p=0.000), and on behavioral intentions on the other hand (β=0.193, t=2.674, p=0.007). Finally, the impact of the website trust on behavioral intentions was supported (β=0.765, t=9.139, p=0.000).

Table 2. Scale measurement properties

	AVE	SP*	RA T***	WT**
SP*	0,675			
RA T***	0,696	,623		
WT**	0,680	,540	,822	
I****	0,462	,619	,777	,833
SP*: Social presence; WT**: Website trust; RA T***: Recommendation agent trust; I****: Intentions				

5 Discussion and Implications

As the usage of embodied recommendation agents increases, it is interesting to understand the effect of their type of voice (human versus synthetic) on perceived social presence, recommendation agent trust, website trust and behavioral intentions. The main findings suggest that both human and synthetic voice are effective in creating a warm social presence and enhancing recommendation agent trust. The impact was, however, greater with human voice (M=0,179 and 0,089) than with synthetic voice (M = -0,212 and -0,105).

In contrast, the voice type had no significant effect on website trust. Some research has shown that the presence of recommendation agent's on website enhances the website's trust (Keeling, McGoldrick, & Beatty, 2010; Qiu & Benbasat, 2005; Wang & Benbasat, 2005, 2008). One possible explanation for these results is that the consumers trust towards the website could be influenced, not only by the presence of the recommendation agent and somehow its voice, but also by many other factors' present, such as website atmosphere, consumer community forum, quality labels.

AMOS results show that all the relationships are significant. Higher social presence perception enhances the recommendation agent trust and engages in more behavioral intentions. As hypothesized, the higher the recommendation agent trust, the more web site trust and intention increases. There was also a positive relationship between the website trust and behavioral intentions. These results support previous research that demonstrated the importance of social factors on the internet and, especially the higher effect of recommendation agent presence on perceived social presence, trust and intentions (Holzwarth, Janiszewski, & Neumann, 2006; Keeling, McGoldrick, & Beatty, 2010; Kohler, Rohm, Ruyter, & Wetzels, 2011; Qiu & Benbasat, 2009b; Wang et al., 2007).

Moreover, the findings of the experiment reinforce the importance of taking the recommendation agent type of voice (human versus synthetic) into account when trying to design a recommendation agent.

From a managerial point of view, these results have several implications and offer implementation guidance to web designers and developers.

Managers and practionners seeking to engage positive intentions through social presence and trust must think of the best way to design their recommendation agent. Indeed, recommendation agents appear on websites as an online brand ambassador. It promotes and conveys its image and must be designed pursuant to a company's values and user expectations. It should inspire confidence and positive perceptions.

However, our findings have also some limitations in areas where future research could be useful. First, this study investigated the impact of voice type on social presence, trust and intentions. It did not consider other characteristics of the recommendation agent, such as facial expressions, gesture, and appearance. So, future research could test the combined effects of these characteristics on user reactions. Future research could also consider a website's atmospherics in order to assess the website's trust. This could help identify which elements have the greatest impact on users.

Finally, further research could consider other kinds of websites and different samples (with students for example). User familiarity with recommendation agent could also be explored.

In conclusion, the recommendation agent voice type is necessary to understand perceived social presence and recommendation agent trust. Based on our study, it is recommended that Website designers take into account the type of voice when designing recommendation agents.

References

1. Burgoon, J., Bonito, J.A., Bengtsson, B., Cederberg, C., Lundeberg, M., Allspach, L.: Interactivity in human-computer interaction, a study of credibility, understanding and influence. Computers in Human Behavior 16, 553–574 (2000)
2. Cassell, J., Bickmore, T.: External Manifestations of Trustworthiness in the Interface. Communications of the ACM 43(12), 50–56 (2000)
3. Cowell, A.J., Stanney, K.M.: Manipulation of non-verbal interaction style and demographic embodiment to increase anthropomorphic computer character credibility. International Journal of Human-Computer Studies 62(2), 281–306 (2005)
4. D'Alessandro, C.: Qu'est ce que la synthèse vocale. Technolangue: le portail des technologies de la langue (2006), `http://www.technolangue.net/article.php3?id_article=275` (in French)
5. D'Alessandro, C.: 33 ans de synthèse de la parole à partir du texte : une promenade sonore (1968-2001). Traitement Automatique des Langues 42(1), 1–29 (2001) (in French)
6. Duffy, B.R.: Anthropomorphism and the social robot. Robotics and Autonomous Systems 42(3-4), 177–190 (2003)
7. Edlund, J., Gustafson, J., Heldner, M., Hjalmarsson, A.: Towards human-like spoken dialogue systems. Speech Communication 50(8-9), 630–645 (2008)
8. Evans, R.E., Kortum, P.: The impact of voice characteristics on user response in an interactive voice response system. Interacting with Computers 22, 606–614 (2010)
9. Fornell, C., Larcker, D.F.: Evaluating Structural Equation Models with Unobservable Variables and Measurement Error. Journal of Marketing Research 18(1), 39–50 (1981)
10. Galanxhi, H., Nah, F.F.-H.: Deception in cyberspace: A comparison of text-only vs. avatar-supported medium. International Journal of Human-Computer Studies 65(9), 770–783 (2007)
11. Grichkovtsova, I., Morel, M., Lacheret, A.: The role of voice quality and prosodic contour in affective speech perception. Speech Communication 54, 414–429 (2012)
12. Gulz, A., Haake, M.: Design of animated pedagogical agents—A look at their look. International Journal of Human-Computer Studies 64(4), 322–339 (2006)

13. Holzwarth, M., Janiszewski, C., Neumann, M.M.: The Influence of Avatars on Online Consumer Shopping Behavior. Journal of Marketing 70, 19–36 (2006)
14. Keeling, K., McGoldrick, P., Beatty, S.: Avatars as salespeople: Communication style, trust, and intentions. Journal of Business Research 63(8), 793–800 (2010)
15. Kline, R.B.: Principles and Practice of Structural Equation Modeling. The Guilford Press, New York (1998)
16. Kohler, C.F., Rohm, A.J., De Ruyter, K., Wetzels, M.: Return on Interactivity: The Impact of Online Agents on Newcomer Adjustment. Journal of Marketing 75, 93–108 (2011)
17. Lee, M.K., Jung, Y., Kim, J., Kim, S.R.: Are physically embodied social agents better than disembodied social agents?: The effects of physical embodiment, tactile interaction, and people's loneliness in human–robot interaction. International Journal of Human-Computer Studies 64, 962–973 (2006)
18. Lee, K.M., Nass, C.: Designing social presence of social actors in human computer interaction. In: Proceedings of the Conference on Human Factors in Computing Systems, CHI 2003, vol. (5), p. 289 (2003)
19. Lemoine, J.-F., Cherif, E.: What is the Best Virtual agent for my Website: An Exploratory Study of the Different Anthropomorphic Characteristics? In: The 33th Annual Conference of the Administrative Sciences Association of Canada, St. John's Newfoundland Canada, June 9-12 (2012)
20. Loiacono, E.T., Djamasbi, S., Kiryazov, T.: Factors that affect visually impaired users' acceptance of audio and music websites. International Journal of Human-Computer Studies 71, 321–334 (2013)
21. McGoldrick, P.J., Keeling, K.A., Beatty, S.F.: A typology of roles for avatars in online retailing. Journal of Marketing Management 24(3-4), 433–461 (2008)
22. Nowak, K.L., Rauh, C.: Choose your "buddy icon" carefully: The influence of avatar androgyny, anthropomorphism and credibility in online interactions. Choose your "buddy icon" carefully: The influence of avatar androgyny, anthropomorphism and credibility in online inte. Computers in Human Behavior 24(4), 1473–1493 (2008)
23. Qiu, L., Benbasat, I.: Online Consumer Trust and Live Help Interfaces: The Effects of Text-to-Speech Voice and Three-Dimensional Avatars. International Journal of Human-Computer Interaction 19(1), 75–94 (2005)
24. Qiu, L., Benbasat, I.: Evaluating Anthropomorphic Product Virtual Agents: A Social Relationship Perspective to Designing Information Systems. Journal of Management Information Systems 25(4), 145–182 (2009a)
25. Qiu, L., Benbasat, I.: Evaluating Anthropomorphic Product Virtual Agents: A Social Relationship Perspective to Designing Information Systems. Journal of Management Information Systems 25(4), 145–182 (2009b)
26. Stern, S.E., Mullennix, J.W., Yaroslavsky, I.: Persuasion and social perception of human vs. synthetic voice across person as source and computer as source conditions. International Journal of Human-Computer Studies 64(1), 43–52 (2006)
27. Stern, S.E., Mullennix, J.W., Dyson, C., Wilson, S.J.: The persuasiveness of synthetic speech versus human speech. Human Factors 41, 588–595 (1999)
28. Steuer, J., Nass, C.I.: Voices, boxes and sources of messages computers and social actors. Human Communication Research 19(4), 504–527 (1993)
29. Wang, L.C., Baker, J., Wagner, J.A., Wakefield, K.: Can a Retail Web Site Be Social? Journal of Marketing 71, 143–157 (2007)
30. Wang, W., Benbasat, I.: Trust in and Adoption of Online Virtual. Journal of Association for Information Systems 6(3), 72–101 (2005)

31. Wang, W., Benbasat, I.: Attributions of Trust in Decision Support Technologies: A Study of Virtual Agents for E-Commerce. Journal of Management Information Systems 24(4), 249–273 (2008)
32. Wongkia, W., Naruedomkul, K., Cercone, N.: i-Math: Automatic math reader for Thai blind and visually impaired students. Computers & Mathematics with Applications 64, 2128–2140 (2012)

Appendix

Construct	Items
Social Presence	I felt a sense of human contact in the agent.
	I felt a sense of personalness in the agent.
	I felt a sense of human warmth in the agent..
	I felt a sense of sociability in the agent.
	I felt a sense of human sensitivity in the agent.
Recommendation agent trust	The agent was competent in recommending insurance products
	The agent performed its role of recommending insurance products very effectively
	Overall, the agent was capable of providing suitable insurance products
	In general, the agent was very knowledgeable about insurance products
	I believe that the agent's dealings with me were in my best interest.
	The agent's dealings with me felt like that it would do its best to help me
	The agent's dealings with me felt like that it was interested in my well-being, not someone else's.
	I believe the agent's recommendations to me were truthful.
	I would characterize the agent's dealings with me as honest.
	The agent appeared to be unbiased.
Website trust	I would believe the information given on this website
	I would trust the payment process on this website
	I would be confident that my order was correct
	I would use the recommendations on this website
Intentions	likely to buy
	likely to recommend
	likely to revisit

Webifying the Computerized Execution of Clinical Practice Guidelines

Tiago Oliveira[1], Pedro Leão[2,3,4], Paulo Novais[1], and José Neves[1]

[1] CCTC/Department of Informatics, University of Minho, Braga, Portugal
[2] School of Health Sciences, University of Minho, Braga, Portugal
[3] Life and Health Sciences Research Institute, Hospital of Braga, Braga, Portugal
[4] ICVS/3B's - PT Government Associate Laboratory, Braga/Guimarães, Portugal
{toliveira,pjon,jneves}@di.uminho.pt,
pedroleao@ecsaude.uminho.pt

Abstract. The means through which Clinical Practice Guidelines are disseminated and become accessible are a crucial factor in their later adoption by health care professionals. Making these guidelines available in Clinical Decision Support Systems renders their application more personal and thus acceptable at the moment of care. Web technologies may play an important role in increasing the reach and dissemination of guidelines, but this promise remains largely unfulfilled. There is a need for a guideline computer model that can accommodate a wide variety of medical knowledge along with a platform for its execution that can be easily used in mobile devices. This work presents the CompGuide framework, a web-based and service-oriented platform for the execution of Computer-Interpretable Guidelines. Its architecture comprises different modules whose interaction enables the interpretation of clinical tasks and the verification of clinical constraints and temporal restrictions of guidelines represented in OWL. It allows remote guideline execution with data centralization, more suitable for a work environment where physicians are mobile and not bound to a machine. The solution presented in this paper encompasses a computer-interpretable guideline model, a web-based framework for guideline execution and an Application Programming Interface for the development of other guideline execution systems.

Keywords: Computer-Interpretable Guidelines, Clinical Decision Support, Framework, Web.

1 Introduction

Clinical Practice Guidelines (CPGs) are systematically developed statements that contain recommendations to assist health care professionals and patients in specific clinical circumstances [1]. Their main goals are to provide patient specific advice, to reduce variations in medical practice and to promote cost containment, through efforts towards the improvement of efficiency and quality. From the appearance of CPGs as information vehicles of medical consensus groups and evidence-based medicine in the

J.B. Perez et al. (eds.), *Trends in Practical Applications of Heterogeneous Multi-agent Systems. The PAAMS Collection*, Advances in Intelligent Systems and Computing 293,
DOI: 10.1007/978-3-319-07476-4_18, © Springer International Publishing Switzerland 2014

late 1970s until the present day their use has become widespread, being regarded by most as useful tools in health care delivery. However, there are some issues that are continuously raised by the medical community [2]. Many believe that paper-based guidelines are difficult to consult at the moment of care. Others think they are an instrument for cookbook medicine and are not convinced that their use leads to better care. Fortunately, the advent of Clinical Decision Support Systems (CDSSs) as tools for information management and providing patient specific recommendations offered a medium through which guideline appliance can become more acceptable, patient tailored and interactive [3]. These are the features that make CPGs implementable during care delivery. As will be seen far ahead, Computer-Interpretable Guideline (CIG) implementations appeared in the early 1990s and have proliferated since then.

In the early 2000s the emergence of t Web 2.0 technologies created yet another opportunity for the evolution of CIGs. It was a change in how the web is perceived by both users and developers, accompanied by the adoption of interaction and participation as fundamental aspects of online activity. The concept of the web as an integrating platform with rich internet applications offers software above the level of a single device. For CIGs this translates into the possibility of their being available anywhere, anytime. This work is an introduction to the *CompGuide* web-based framework for CIG execution.

The article is organized as follows. Section 2 provides insights on current CIG development and the relevant work done in the field. Section 3 gives a brief explanation of the model used for CIG representation. The framework is presented in Section 4. Finally, Section 5 provides some conclusions about the work done so far and future considerations for the work ahead.

2 Relevant Work

Since the 1990s many researchers have developed and proposed CIG models to represent guidelines in a computer-interpretable language. Arguably, the most relevant are Arden Syntax [4], the Guideline Interchange Format (GLIF) [5], Asbru [6], PROforma [7] and the Standards-based Active Guideline Environment (SAGE) [8]. Apart from Arden Syntax, which is a model specifically for encoding small fragments of clinical knowledge in the form of rules, the majority of models employ some kind of task-oriented network to represent and display guideline knowledge. The Task Network Model (TNM) appears to be the one that best fits the information conveyed by guidelines, allowing for a clear separation between procedural knowledge, i.e. the relative order of recommendations, and medical knowledge. A model is usually accompanied with an execution engine which is responsible for interpreting the clinical constraints and temporal constraints placed on tasks. Tools such as Arezzo™ (for PROforma), the Digital Electronic Guideline Library (DeGeL) (for Asbru), the Guideline Execution Engine (GLEE) (for GLIF3) and SAGEDesktop are used to provide recommendations interactively and store execution traces of guidelines represented in their respective models. Normally, the execution engines comprise a desktop client application that remotely connects to a server which provides a guideline repository. This tendency for the development of desktop applications is evident in Isern and Moreno's review work [9] on computer-based execution of CPGs. Despite the virtues of this type of structure and desktop applications, one is forced to consider that CIG development is not taking advantage of the web's full capabilities to increase the reach and availability of CPGs, their level of dissemination, and their scrutiny. A comparative analysis of current approaches to CIGs was done in previous work [10], where a comparative table regarding their main features is provided.

In a recent review article [11], Peleg indicated the development of ubiquitous CIG-based systems as an emergent trend. Through web-based and smartphone/tablet interfaces, these systems should be able to provide ever accessible clinical decision support. There are projects that have already started to develop their CIG solutions under the concept of guidelines as services. The work presented herein assimilates these ideas, and implements a Service Oriented Architecture (SOA) to make available the functionalities that enable the execution of CIGs. It comprises a guideline representation model, a server application that provides a set of guideline-related services and a web interface that makes full use of these services.

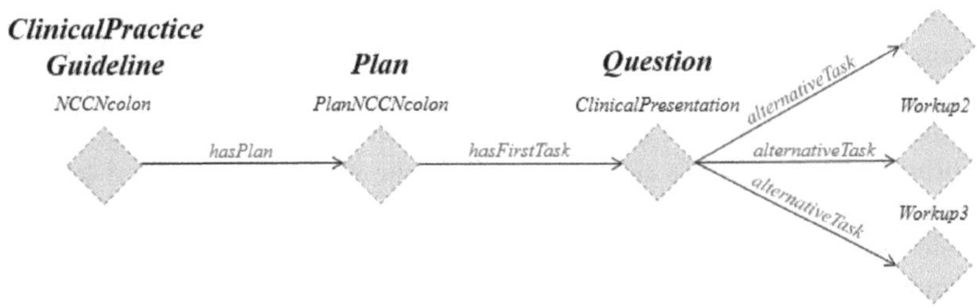

Fig. 1. Initial formalization of the NCCN Guideline for Colon Cancer in CompGuide ontology. A guideline is represented as a network of individuals connected by object properties.

3 Computer-Interpretable Guideline Model

The *CompGuide* ontology model for CIGs, which is used in this work, is expressed in OWL-Description Logic (OWL-DL) version 2 [13], and it has been previously presented in [12]. Protégé was used as an ontology development tool and in the creation of guidelines. In the ontology, a CPG is represented as an individual of the class *ClinicalPracticeGuideline*, which has a set of data properties to express administrative information and object properties to connect it to individuals of other classes. This set up may be seen in Fig. 1 which is the initial formalization of the National Comprehensive Cancer Network (NCCN) Guideline for Colon Cancer according to the ontology. The TNM is implemented in the form of four classes of tasks: *Plan*, a collection of tasks containing any number of other tasks, including other plans; *Action*, a task performed by a health care agent, namely a clinical procedure, a clinical exam, a medication recommendation or a non-medication recommendation; *Question*, an enquiry task to obtain information about the patient; and *Decision*, a reasoning task about the state of a patient which implies the choice between two or more options, yielding a conclusion which is then used to update the state of the patient.

Similarly to what is expressed in Fig. 1, different tasks are represented by individuals of the above-mentioned classes. Being a workflow representation, there have to be control structures to define the relative order between tasks. A guideline has a main *Plan* which contains all the tasks. The individual corresponding to this *Plan* has, in turn, an object property that points to its first task. Then, the previous tasks always indicate those which follow. It is possible to define sequential tasks, tasks which should be executed at the same time (parallel tasks) and alternatives in the guideline workflow (alternative tasks). For parallel tasks it is also possible to define synchronization tasks where the workflow reunifies after splitting.

A CIG model requires the appropriate constructs for the definition of clinical constraints. In this regard, *CompGuide* offers the possibility to define *TriggerConditions* which are used to specify the terms regarding the patient state that dictate the choice of an alternative task. Other clinical constraint classes are *PreCondition*, which specifies the conditions that must be met before executing a task, and *Outcome*, the expected result of a task in terms of the alterations it produced in the patient state. It is also possible to define, in *Decisions*, the options to choose from and rules associated with them.

Temporal restrictions are also an important element of medical algorithms. Thus, *CompGuide* provides *Periodicity* and *Duration* classes. The former may be used to express from when to when a task should be executed and/or its number of repetitions. Through *Periodicity* it is also possible to define stop conditions for a cyclic task and, in the event of these stop conditions holding true, the task the guideline execution should move to, which is a stop condition task. The *Duration* indicates how long a task should last.

4 CompGuide Framework

The CompGuide framework was created in order to provide tools capable of automating the functionalities offered by the expressiveness of the ontology, to automatically process workflow control structures, clinical constraints and temporal restrictions. The following briefly introduces the tools and technologies used in the development, explains the framework's architecture and addresses the most significant aspects regarding guideline execution.

4.1 Tools and Technologies

The tools and technologies which were used were chosen by taking into consideration the requirements of the application's architecture, namely its strong web component, as mentioned in Section 2. Accordingly, web services were used as the preferential form of communication. Their usage offers expandability and the possibility to improve services without compromising others. For their lightness and ease of access, Representational State Transfer (REST) Web Services were the chosen service model.

To stay aligned with the overall goal, the framework was developed as a Web Application Programming Interface (Web API) to enhance multi-platform implementation capabilities.

The access to guidelines in the OWL ontology file is done through the Java OWL-API, developed and maintained at the University of Manchester. It provides an easy way to create, manipulate and serialize OWL ontologies. The web application that serves as user interface was developed using Java Server Pages (JSP). The data exchanged with the server is in JavaScript Object Notation (JSON), a text-based open standard for data interchange. It offers simple parsing and, at the same time, is compact when transferring small amounts of data.

The user, patient and execution information is stored in a MySQL relational database. The applications are deployed in a JBoss application server.

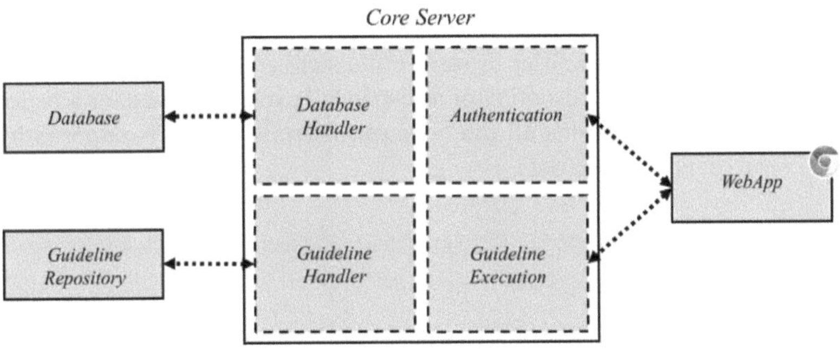

Fig. 2. Architecture of the *CompGuide* framework with the Core Server connecting the data storage components to the web application

4.2 Architecture

The framework's architecture is depicted in Fig. 2. It is composed of four separate main components. The communication between the main applications, in particular with the web application, is enabled by web services. The data storage components feature a *Database* model to store user, patient and execution information, and the *Guideline Repository* which is an OWL file in order to store CPGs. The *Core Server* uses JBoss to provide the services required for guideline execution to external applications. It is responsible for all the database manipulation and guideline execution control. It takes advantage of JBoss's features by implementing session controllers, user authentication and publication of web services (accessible through HTTP methods). Within the *Core Server* there are different components which work under the web service layer, being seamless for the user interface applications. Said components comprise the following modules:

- *Authentication*: the module provides both authentication and authorization. Two types of users are defined: *admin* and *user*. While a *user* has full access to manipulate user and patient information (add, remove and edit) and execute guidelines, an *admin* has access only to the guideline execution functionality. The authentication process is done through a specific web service which sends a token as response. This token is then used to request the other services until expiring and ensures that the level of access of the user is respected. The token verification is handled by a filter.

- *Guideline Handler*: this module does the retrieval of guidelines from the *Guideline Repository*. It includes a set of functions to fetch the information for each class on the guideline ontology. In big hand coded guidelines there are often mistakes, such as missing components and properties, hence, a syntactic verification tool was created so as to guarantee the validity of a guideline in this regard.
- *Guideline Execution*: guideline execution is done entirely in the *Core Server* by this module. If offers two distinct services, both responsible for computing the next task in the guideline algorithm as well as providing the information featured in it. The difference between the two is that one of them allows to skip a task if the physician desires to do so, while the other does not. When the task is executed, the module stores this event in order to restore the state of the execution to previous moments. This enables the resuming of a suspended execution trace for a patient.
- *Database Handler*: it contains all the updates, inserts and selects required by the *Core Server* to handle stored data.

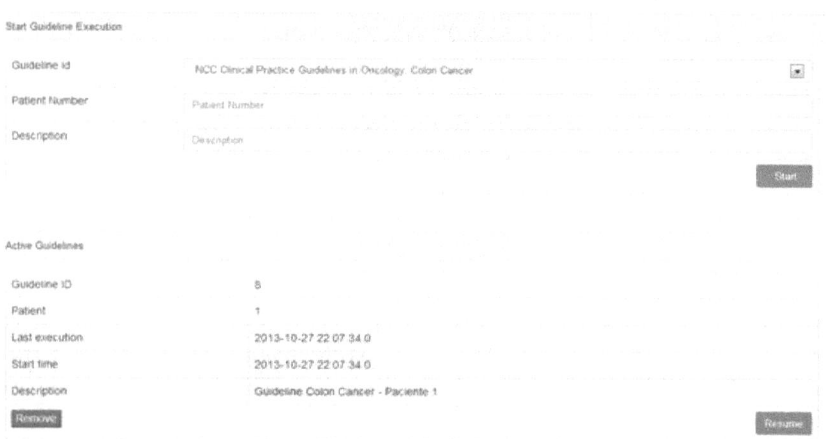

Fig. 3. Interface of the guideline management suite showing the active guidelines with the option to resume their execution and the new guideline ready to be deployed

The implemented *Database* model stores all the user and patient information combined with the information produced by the interface application. This allows for the mapping of user and patient records to every guideline execution instance. The structure is based on an OpenEHR archetype model [13]. The information is saved as *observations* (relevant states of the patient retrieved through *Question* tasks) and *actions* (exams, medication, non-medication recommendations and procedures).

The *Web App* aims at collecting information from and delivering information to the user. It provides a *control panel* which enables the editing of the personal information of users and patients, and a *management suite* (as shown in Fig. 3) that offers the possibility to start and resume the execution of guidelines. When starting a new guideline, the user must specify the patient to whom the guideline will be applied. The interface for the guideline execution covers all the tasks mentioned in Section 3, with tasks being presented as they are computed by the execution engine.

4.3 Aspects of Guideline Execution

In CIG execution, there are essentially three types of verification that execution engines must perform. They are task ordering verifications, clinical constraint verifications and temporal restriction verifications.

The model allows the representation of different modalities of task performing, from sequential tasks and parallel tasks to alternative tasks. Mixing these types together, which may occur in real guideline algorithms, significantly increases the complexity of the programming necessary to handle these situations. As an example, a parallel task may be followed by another set of parallel tasks or alternative tasks which, in turn, can also be followed by other such tasks. This raises issues in keeping their synchronization points in check. To tackle them, a *task controller* was developed for storing information at three levels: the task, the plan, the parallel tasks and the alternative tasks. It ensures that the execution engine follows a plan and that all parallel and alternative tasks are synchronized. The patient state is built through *Question* tasks. The information is collected and stored as an *observation*. Whenever it is necessary to check trigger conditions for alternative tasks or validate rules in *Decision* tasks, the values of the clinical parameters contained in them are searched in *observations*. A *PreCondition* is verified before proposing a task and an *Outcome* after performing it. *StopConditions* are verified at every iteration of a periodic task. Every task entry in the *Database* has a timestamp. This is used to store the moment when a specific task is performed and to control the different temporal restrictions. When a temporal restriction is met, a warning is issued to the user.

5 Conclusions and Future Work

Leaving the heavier processing, such as task computation, to the server proves to be an effective implementation of CIG execution, removing the need to exchange big chunks of information. Comparing with existing systems, the hereby presented architecture offers some advantages, in particular it provides an API to access an OWL CPG ontology that other developers can use in their own application, it is easy to access its functionalities given the service-oriented architecture, and it offers the possibility to easily develop other user-interfaces. It allows remote guideline execution with data centralization, more suitable for a work environment where physicians are mobile and not bound to a machine. Nevertheless the usefulness of such a framework, its development is still at an early stage. The *Core Server* requires more modules that implement functionalities that are missing such as guideline creation and terminology services. The integration of Unified Medical Language System (UMLS) features is within the scope of the project, as is the mapping of the information produced by guideline execution to standards of medical information exchange, such as the Health

Level Seven (HL7) Clinical Document Architecture. The original contributions of this work are: a computer-interpretable guideline model which can be used for any type of guideline and does not require knowledge on any programming languages; a web-based platform for the execution of clinical practice guidelines in a simple and intuitive way; and an API that developers can use in their own implementation

In the long run, the information stored by the system may be useful in order to assess how guidelines cope with situations and if physicians conform to them.

Acknowledgements. This work is part-funded by ERDF - European Regional Development Fund through the COMPETE Programme (operational programme for competitiveness) and by National Funds through the FCT – Fundação para a Ciência e a Tecnologia (Portuguese Foundation for Science and Technology) within project FCOMP-01-0124-FEDER-028980 (PTDC/EEI-SII/1386/2012). The work of Tiago Oliveira is supported by doctoral grant by FCT (SFRH/BD/85291/2012).

References

[1] Field, M.J., Lohr, K.: Guidelines for Clinical Practice: From Development to Use. The National Academy Press, Washington DC (1992)

[2] Woolf, S.H., Grol, R., Hutchinson, A., Eccles, M., Grimshaw, J.: Potential benefits, limitations, and harms of clinical guidelines. BMJ Br. Med. J. 318(7182), 527–530 (1999)

[3] Latoszek-Berendsen, A., Tange, H., van den Herik, H.J., Hasman, A.: From clinical practice guidelines to computer-interpretable guidelines. A literature overview. Methods Inf. Med. 49(6), 550–570 (2010)

[4] Samwald, M., Fehre, K., de Bruin, J., Adlassnig, K.-P.: The Arden Syntax standard for clinical decision support: Experiences and directions. J. Biomed. Inform. (February 2012)

[5] Peleg, M., Boxwala, A.A., Ogunyemi, O., Zeng, Q., Tu, S., Lacson, R., Bernstam, E., Ash, N., Mork, P., Ohno-Machado, L., et al.: GLIF3: the evolution of a guideline representation format. In: Proceedings of the AMIA Symposium, p. 645 (2000)

[6] Balser, M., Duelli, C., Reif, W.: Formal Semantics of Asbru – An Overview. Science (80) (2002)

[7] Vier, E., Fox, J., Johns, N., Lyons, C., Rahmanzadeh, A., Wilson, P.: PROforma: systems. Comput. Methods Programs Biomed. 2607(97) (1997)

[8] Ram, P., Berg, D., Tu, S., Mansfield, G., Ye, Q., Abarbanel, R., Beard, N.: Executing clinical practice guidelines using the SAGE execution engine. Stud. Health Technol. Inform. 107(pt. 1), 251–255 (2004)

[9] Isern, D., Moreno, A.: Computer-based execution of clinical guidelines: a review. Int. J. Med. Inform. 77(12), 787–808 (2008)

[10] Oliveira, T., Novais, P., Neves, J.: Development and implementation of clinical guidelines: An artificial intelligence perspective. Artif. Intell. Rev. (March 2013)

[11] Peleg, M.: Computer-interpretable clinical guidelines: A methodological review. J. Biomed. Inform. 46(4), 744–763 (2013)

[12] Oliveira, T., Novais, P., Neves, J.: Representation of Clinical Practice Guideline Components in OWL. In: Pérez, J.B., et al. (eds.) Trends in Pract. Appl. of Agents & Multiagent Syst. AISC, vol. 221, pp. 77–85. Springer, Heidelberg (2013)

[13] Gutiérrez, P.: OpenEHR-Gen Framework (2010)

The New Competencies Development Level Expertise Method within Intelligent Automated Educational Systems

Viktor A. Uglev and Valery A. Ustinov

Center for Applied Research of Siberian Federal University
Zheleznogorsk, 662971 Russia
uglev-v@yandex.ru

Abstract. The problem of the competencies development level assessing is considered while working with automated educational systems. It is proposed to use expert approach to mediated assessing of the competencies development level within intelligent educational systems. A technique for obtaining numerical estimates of competences according to the result of the test materials passing (tests solutions used as an example), also their normalization with assessment standards identifying are described. The problem of individualization in learning based on a set of software agents of teacher, student, and tutor is considered.

Keywords: automated training systems, automated educational systems, expert measurement, competence development, computer test, agents.

1 Introduction

Intelligent automated educational systems have ensured the presenting of educational material process, its control and the managing decisions elaboration. Since a student man as an object of management is a complex and weakly formalized system that special attention is given to knowledge control process and level of competence development (LCD) control process. If there is maximum degree of automation of the individual learning process, the only way to achieve an adequate assessment of LCD possible is to have the appropriate methodology and software. Let's consider the problem of diagnosing competencies within intelligent automated educational/ training systems (IAES), based on the expertise and conventions of cybernetics (Wiener 1948).

The need to assess LCD in the automated mode is a consequence of the transition to a new paradigm of learning, based on the information processes and principles of individualization (Howard et al. 2004, Schejbal 2012). The traditional approach to the diagnosis of LCD is based on the application of psychological and qualimetric tests, those are characterized by high detachment degree from the educational process. While implementing automated learning a knowledge assessment methodology often substitutes the methodology of LCD assessing (e.g. see Zvonnikov and Chelyshkova

J.B. Perez et al. (eds.), *Trends in Practical Applications of Heterogeneous Multi-agent Systems. The PAAMS Collection*, Advances in Intelligent Systems and Computing 293,
DOI: 10.1007/978-3-319-07476-4_19, © Springer International Publishing Switzerland 2014

2009) or an adaptation of private highly specialized techniques takes place (for example, see Argimon-Pallàs et al. 2009). A number of standardized approaches, such as, VET PISA (Rauner et al. 2013) or (Woloszynsky and Kurzynski 2009), also have difficulties in IAES-based implementation and often rely on existing specialized test items databases. Obviously the method of measuring knowledge and measure LCD should not be identical and require special elaboration, testing and tools (especially with regard to computer tests as the most popular means of controlling via IAES, for example, Brusilovsky 2003, Uglev 2011, Shavelson 2013).

An expert man while facing the task of diagnosis LCD copes with this task by pushing for each competence hypothesis as it develops in the individual, and then checking the level of all competence through the questions/tasks, taking into account uncertainty. The student's goals, structure of the course, the current and previous learning situation are taking into account, i.e. afferent synthesis occurs (Anokhin 1974). Consequently, it is necessary to simulate the logic of reasoning expert forming a special knowledge base which accompanies test materials IAES.

In the theory of artificial intelligence and decision-making iterative hypothesis testing is one of the key challenges that is reflected in the implementation of a set of individual techniques of expert evaluation with uncertainty: Bayes approach, Naylor approach, Shortliff approach, Dempster approach and others approaches are most well-known (Jackson 1998). We apply expert evaluation to the problem of estimating the LCD in the framework of IAES.

2 Competencies Development Level Expertise Method

2.1 Problem Definition

Let d_i is didactic unit of a set of elements of the course with a sequence number i; T_i is a questions subset out of a test items bank of the course relevant topic, A is a set of developing competencies, and E is meta-information about the e-course. Confining ourselves to investigational question, let's consider E as only the parameters of test materials data (difficulty measure, correctness of responses measure, and controlling and training materials communication measure). If set X_t is an IAES protocol about passing the test at the t time, the assessment of knowledge may be presented as evaluation vector DR for each piece of course i. If relation d_i - T_i can be described as one to many, then each question out of T_i can have multiple connections with individual elements A (Fig. 1). Each test task distractor (answer variant), which is included in the alternate test, contributes to dr_i assessment, thereby allowing to extract data out of X_t to generate assessment of subject knowledge. There will be missed issue of determining the answer correctness to a single test task (methods may be various).

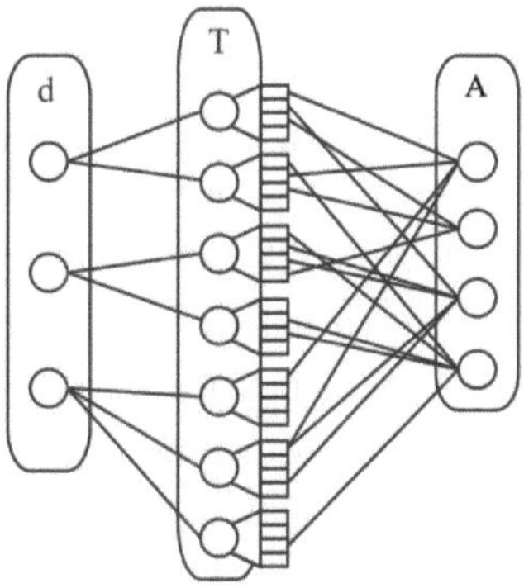

Fig. 1. Link between didactic material and list of competencies

Let's consider the LCD assessment at present. So far as a basis was taken from the expert assessment method, it is obvious that it would require the development of the following aspects:

- Knowledge Base;
- Solver algorithm (method);
- Standards formation algorithm (method).

Let us examine them sequentially, highlighting the specificity of each aspect..

2.2 Expert Assessments Knowledge Base

Expertise in the problem of LCD estimating are both test materials T_i and meta-information E. Controlling and measuring material must possess a number of properties:

- assessment complexity - subject knowledge assessment and LCD assessment processes integration;
- LCD control latency - student have to know about subject knowledge control, and LCD is estimated indirectly;
- hypotheses testing parallelism - the number of jobs to check shall be minimized by combining in one job showcasing several manifestations LCD (if it is possible);
- metricity - the possibility of prior estimation scale introducing which allows to assess the results dynamics in the case of further processing through IAES core (reference type synthesis).

If we talk about organization of control, for example, in the form of test tasks, then the form of the question and distractors should be formulated in such a way that allows the students to have a choice according to their preferences, instead of true or false choice (for a subset out of T_i). Obviously, an expert of subject should form these controlling and measuring materials, basing on a predetermined set of competencies.

Actually the knowledge base in relation to e-learning course is a piece of meta-information (Uglev 2008). Therefore expert assessment of a bunch of elements T_i and A become integral part of E, as shown in Fig. 1. Let each task distractor (answer variant) can be connected with any of developed competences k through confirming or refuting that student has been acquiring that competence with the weight q_{ujk} of the domain [0, 1], where u is sequence number of the question, j is sequence number of distractor, which is connected to competence a_k. Then the task of LCD assessing (quantitative values determination of vector AR) can be reduced to the hypothesis testing of problem of the student's LCD development measure, based on the evidence totality (responses data out of X_t) which consists of evidences having expert estimates q_{ujk}. During IAES core implementation there are two possibilities to store expert assessments values: as an adjacency matrix and as a table in a relational database composed of controlling and measuring materials. For example, if value q_{ujk} is equal to zero it can be taken as negative evidence about the maturity of individual competence, if value q_{ujk} is equal to 1 it can be taken as evidence for confirmation, and if q_{ujk} value is equal to 0.5 it can be taken as equivalence of the main and opposite hypotheses.

2.3 Competence Development Assessment

On the basis of the knowledge base can produce a numerical estimate of the LCD. To do this for each of k competencies we should perform the following steps:

1. To sample student's answers to those test items that were defined in the knowledge base through prior estimates of q_{ujk};
2. To verify the hypothesis that the current competence is developed by calculating measure of belief value, called MB, through the left formula of (1).
3. To verify the hypothesis that the current competence is developed by calculating measure of disbelief value, called MD, through the right formula of (1).

$$MB[a_k, x] = \frac{p(a_k / x) - p(a_k)}{1 - p(a_k)} \quad \text{и} \quad MD[a_k, x] = \frac{p(a_k) - p(a_k / x)}{p(a_k)}, \tag{1}$$

where x is a quantitative measure of one of the factors (test items) for each competency of A.

4. To get overall certainty coefficient Cf about developmental level of measured competence by the formula (2), proposed in (Buchanan and Shortliffe 1984) for composite hypotheses testing:

$$Cf(a_k, q_a \wedge q_{\bar{a}}) = MB(a_k, q_a) - MD(a_k, q_{\bar{a}}), \tag{2}$$

where $a_k \in A$, whereas q_a and $q_{\bar{a}}$ are confirmation evidence (q_{ujk} >0.5) and negative evidence (q_{ujk} <0.5) for hypothesis of level of competences development as a result of work with IAES control subsystem.

5. To present results for each competency in numerical and graphical forms (for example, making use of icon plot).

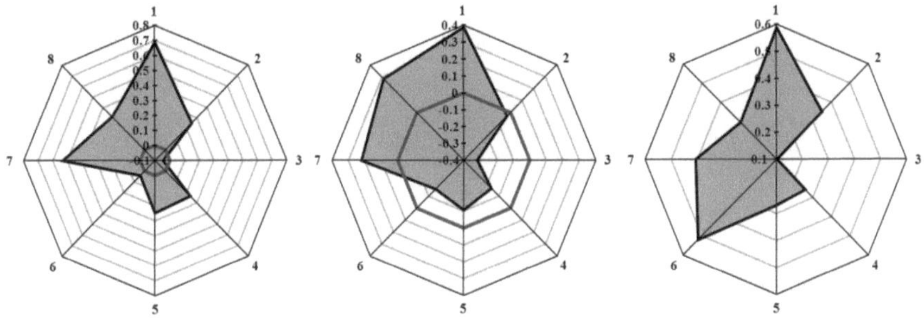

Fig. 2. Examples of LCD set evaluation results with IAES tools

Thus as a result of the test solution in X_t, along with DR estimates a number of certainty coefficients will be obtained corresponding to the degree of each competence manifestation (evaluation vector AR). For example, icon plots, as in Fig. 2, show the results of three students' LCD set evaluation after students passed the on-receipt inspection test of 60 questions, where zero value of line is the boundary of IAES uncertainty about developed or underdeveloped student's competence.

While implementing such a LCD diagnostic algorithm, student's achievements assessment task appears. On what scale is the resulting certainty coefficient in the development of competence $Cf(a_k)$ measured? And how can it be assessed? To answer these questions we consider the method of standards forming and the evaluations results normalization.

2.4 Reference Model

Acquisition of test reference parameters and obtained with formula (2) results normalization are pursue following objectives:

- Acquisition of assessment of diagnostic test potential in the context of each competence;
- Determination of the competency development degree for a certain scale (the task of tracking the results dynamics);
- Comparison of different students' results, provided that they solved the same tests set.

The last objective is important for students assessment where the test material is the same for all, and its variability is achieved by test items shuffling in the test sample and by distractors shuffling inside each of test items.

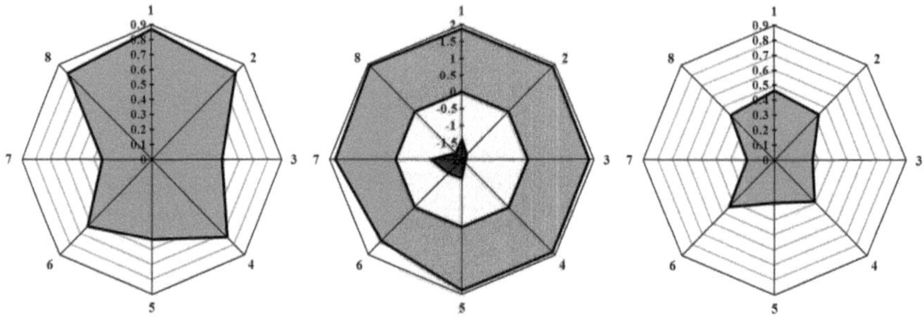

1) 2) 3)

Fig. 3. The initial evaluation results (1), reference model (2), normalized estimates (3)

The reference model parameters calculation and the results normalization for each of k scales involve calculation of the worst and the best of competency evaluations possible results on a particular test sample according to the formula (2). For example, it, as in Fig. 3-1, is shown the results of student's LCD set primary evaluation after student decided 45 questions for checking knowledge of the subject and 8 competences. Reference profile determination is a totality of indicators (vector m_k, which is schematically shown in Fig. 3-2), which are subject to next normalization to a single scale in interval $[-s\,;s\,]$ according to the formula (3) (Fig. 3-3 when $s=1$).

$$m_k = AR_k * s/(\max_k(X_t)). \tag{3}$$

LCD values which are obtained after normalization may will be compared to the previous values and may will not be compared to the values of other competencies later.

3 IAES Work Individualization and Agent-Based Approach

IAES functionality improving (especially used in WANs) is aimed at satisfaction of the society's educational order which requires individualization in learning in the information age (Kravets 2011). An individualized approach implementation can be based on person's goals (including competency preferences) in the following aspects: the didactic material composition, the trajectory of learning, individual control standards, flexible dialogue in natural language interaction form (Mashbits 1988, Uglev 2010). Obviously, the intelligent training management requires that IAES should identify and take into account the student's preferences and the purposes of the e-learning course developers simultaneously. In fact, it is necessary to overcome the conflict of interests.

From the standpoint of IAES composition it is the conflict between the student's model and the teacher's model which do not have the coincident goals while working with the system. Then they can be represented as agents those need to reach compromise. Tutor can solve this conflict in traditional individualized learning who acts as an arbitrator of interests conflict: his purpose is to find a reasonable compromise between the curriculum requirements and student's preferences. Thus, it is efficiently to enter a tutor agent in IAES, which applies agent-based architecture, the tutor agent helps to reveal and form the individual curriculum and trajectory of its mastering.

Note that the actions of all three agents involved in settling of interests conflict are tied essentially to the possibility of IAES to determine LCD, and agents have to take into account those while they are making decisions in the process of forming an effective individual educational trajectory (Uglev 2012).

4 Results and Conclusion

The proposed approach to assessing the level of competencies development is focused on modern intelligent educational systems. As an example, let's give some experimental results obtained during training group of equipment spacecraft operators: based on program-mathematical model within IAOS learning was accompanied with periodic tests in the training and control regimes. In Fig. 4. the dynamics of average students' ratings in the group of competencies is shown: the abscissa scale designates the number of intermediate certification (once a quarter), y-axis designates the scale of confidence in the well-developed competence. Not only the positive dynamics of subject knowledge learning was obtained, but also an LCD increase was recorded, which was recorded indirectly.

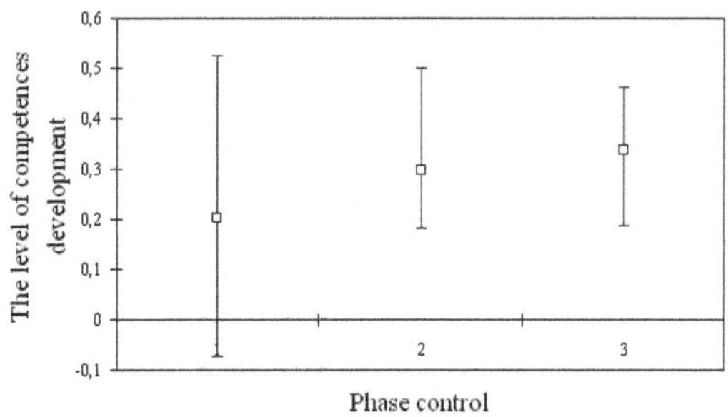

Fig. 4. Summarized LCD dynamics of the three intermediate certifications of group of students

These results confirm the hypothesis of studies that the competence assess is available in the automated mode. At the same time the estimates source may be not only a set of tests, but also exercises in virtual laboratories, in robotic educational stands (Khnykin et al. 2011), and in program-mathematical models (Uglev 2013).

Acknowledgements. This research is supported by the Ministry of Education and Science of the Russian Federation in Siberian Federal University (contract № 02.G25.31.0041).

Also thank Grachevu Karinu for helping to develop visualization module.

References

1. Anokhin, P.: Biology and Neurophysiology of the Conditioned Reflex and Its Role in Adaptive Behavior. Pergamon, Oxford (1974)
2. Argimon-Pallàs, J., et al.: Study protocol of psychometric properties of the Spanish translation of a competence test in evidence based practice: The Festo test. BMC Health Services Research (2009), doi:10.1186/1472-6963-9-37
3. Brusilovsky, P.: Adaptive and Intelligent Web-based Educational Systems. International Jornal of AI in Education 13, 156–169 (2003)
4. Buchanan, B., Shortliffe, E.: Rule-based Expert Systems. Addison-Wesley, New York (1984)
5. Howard, C., Schenk, K., Discenza, R.: Distance Learning and University Effectiveness: Changing Educational Paradigms for Online Learning, London (2004)
6. Jackson, P.: Introduction to Expert Systems. Addison-Wesley, New York (1998)
7. Khnykin, A., Laletin, N., Uglev, V.: Applying Zheleznogorsk Robotics for Learning Children with Disabilities. In: Bravo, J., Hervás, R., Villarreal, V. (eds.) IWAAL 2011. LNCS, vol. 6693, pp. 167–171. Springer, Heidelberg (2011)
8. Kravets, O.: The modern pedagogical technologies analysis of automated individualization training. Psychology, Sociology, Pedagogic 4, 14–17 (2011) (in Russian)
9. Mashbits, E.: Psycho-pedagogical problems of education computerization. Moscow (1988) (in Russian)
10. Rauner, F., et al.: Measuring Professional Competence. COMET (2013), doi:10.1007/978-94-007-4725-8_1
11. Schejbal, D.: In Search a New Paradigm of Higher Education. Innovative Higher Education 37, 373–386 (2012), doi:10.1007/s10755-012-9218-z
12. Shavelson, R.: An Approach to Testing & Modeling Competence. Professional and Vet Learning (2013), doi:10.1007/978-94-6091-867-4_3
13. Uglev, V.: On the specificity of individualization of training in Automated Training Systems. Philosophy of Education 2, 68–74 (2010) (in Russian)
14. Uglev, V.: Intellectual Control Algorithm Interaction improvement by the Users Education Process of the Automation Education Systems. In: International Siberian Conference on Control and Communications (2011), doi:10.1109/SIBCON.2011.6072615
15. Uglev, V.: The Cognitive Maps of Knowledge Diagnosis. Open and Distance Learning 48, 17–23 (2012) (in Russian)
16. Uglev, V.: The specificity of operator training on the basis of processfocused on model Automated Educational Systems. Questions of modern science and practice. Vernadsky University 48, 54–58 (2013) (in Russian)
17. Uglev, V., Samrina, F.: Using of possibilities in Learning Tests for individualization of displaying material in Electronic Education Courses. Modern Techniques and Technologies (2008), doi:10.1109/SPCMTT.2008.4897509
18. Wiener, N.: Cybernetics or Control and Communication in the Animal and the Machine, New York (1948)
19. Woloszynski, T., Kurzynski, M.: On a New Measure of Classifier Competence in the Feature Space. In: Kurzynski, M., Wozniak, M. (eds.) Computer Recognition Systems 3. AISC, vol. 57, pp. 285–292. Springer, Heidelberg (2009)
20. Zvonnikov, V., Chelyshkova, M.: Quality control training for certification: competence approach, Moscow (2009) (in Russian)

Cooperative Robots Used for the Learning Process in the Cooperative Work

Yoshua Haim Ovadiah[1], Gabriel Muñoz Samboni[2], and John Páez Rodríguez[3]

[1] Universidad Distrital Francisco José de Caldas, Lic. Physic
[2] Universidad del Cauca, Systems Engineer
[3] Universidad Distrital Francisco José de Caldas, Faculty of Science and Education
Bogotá, Colombia
http://www.udistrital.edu.co

Abstract. This article describes the design and building process of a Learning Virtual Object that uses the Multiagent Systems theory to promote the thinking about advantages of cooperative work while solving problems. The software context is the environment care and for its execution users establish the cooperative conditions of three different robot groups responsible for collecting, classifying and storing three types of recyclable materials. Throughout the execution, the user evaluates if the established cooperative conditions are accurate to the correct task development. The software has been designed using Eclipse Kepler as IDE, *Processing* version 2.1 and the library AI for 2D Games and G4P to manage the robots states and the user interface respectively. The implementation results prove the software accuracy to learn how to cooperate in daily basis tasks through the cooperative robots programming, besides a conceptual change about cooperative concept that is evident in users, the team works the advantages and the distribute cognition principles for the cooperative tasks development.

Keywords: Multiagent Systems, Robotics, Cooperative Work, Virtual Learning Processes Environments.

1 Introduction

The National Education Ministry of Colombia has proposed policies to promote the social skills development to improve the coexistence and the participation in the different social processes in the country[2]. Among the reasons, which motivate the policies advance, there are the ethnic and cultural diversities, the social cohesion enforced processes and the challenges that globalization process brings concerning about the cooperative work dynamics due to the interdisciplinary nature complexity in the social scientific and technologic problems and the way they are currently solved. Unfortunately, the evidence of these learning processes for the productive and social fields still emerges [1],[3], [19]. That is why, it is important to continue the innovative strategies development to reinforce the cooperative work skills that increase the country's growing expectations.

J.B. Perez et al. (eds.), *Trends in Practical Applications of Heterogeneous Multi-agent Systems. The PAAMS Collection*, Advances in Intelligent Systems and Computing 293,
DOI: 10.1007/978-3-319-07476-4_20,

Some research [4],[5] presents the cooperative work as a strategy to solve problems in a creative and efficient way because of the following reasons: several people participate with different abilities, self organization group capability, collective intelligence, distributed cognition, rational resources use, among others. However, the implementation of this processes is not simple due to the fact that it requires an appropriate knowledge to integrate the work groups, define its structures, roles, identify the assertive leaderships processes, communication, conflicts resolution, making decisions and working out in the collective culture. All these framed an emotional collective and individual intelligence which facilitates the social relationships, stimulates the confidence, promotes the individual and social commitment, brings cohesion and efficacy and avoids lack of motivation, laziness, free-ride and responsibilities eluding.

The context below produces the challenge of making the educative policies intentions come true and that individuals understand, in a educational way, the consequences in the variation of the cooperative aspects during the development of cooperative activities. Through this learning process they identify their behavior while they are working in groups, and with the facts they can positively influence to improve cooperative tasks. The way they impact their environment and how they help to build a better country, which in the current conditions of globalization should face the economic social and cultural challenges in a collective method.

Several authors [16] introduce the simulation advantages as a process of cognitive mediation which allows to imitate, comprehend, precede and theoretically verify the different systems. Some authors talk about simulation as a method that involves inductive thinking and depends on deductive process to get close to knowledge[18]. When talking about Multiagents Systems, it is possible to say that this simulation combines the abilities of two or more agents to solve a problem, which, due to its complexity, cannot be solved by one agent alone. Combination is supported by different organizational approaches that separate the problem according to its complexity and the agents design conditions, this determines the coordination, cooperation, communication and the conflict solving processes. Another methodology is inspired on swarms[6], where the behavioral explanations are supported on the quantity of individuals. When they are in a community and we add their individual behaviors they look like a *super-organism* with evidence of intelligence and distributed knowledge.

In robotics, agents are either abstract or physical. Agents are characterized as autonomous, with clear objectives, resources, states and actions identities which are elements needed to navigate and knocked into shape environments by passive or active objects. They perform in a reactive, deliberative or in a combined effort to achieve the goals according to certain beliefs, desires and intentions[7],[8].

Now, robotics, as a didactic strategy[20][21], has been growing during the last few years mainly by the *motivation*. This can be philosophically explained in aspects such as: The man desire to reproduce the organic and to reach the power to conquer the natural things; The need to transfer God's order to the machine; Gain freedom and not to depend on natural forces to skip evolutionary steps; to create artificial systems which challenge the thinking and get manifestations of intelligence; to create new worlds and reach an artificial living way, among others. It is also related to the *animism*, which is a trend that believes inanimate objects are gifted with personality, soul and life [10],[11],[17]. Bringing up the mentioned ideas: simulation as a cognitive mediation tool, multiagent systems as a discipline which shapes human being collective behavior, and robotics as a motivational tool in the learning process. The contributions are: Real time feedback of the different cooperative acts that result in consequences while solving a problem; The process from concrete to abstract is when the individual identifies a concept of cooperation and demonstrates it in the software; Raising the motivation during the learning process and also understanding the cooperation concept from the systematic and complex perspective.

2 Software Development

The figure 1 describes the interface, compound by a two-dimensional *environment* where there are differentials mobile *robots (yellow, blue, green)* navigating and interacting according to the cooperative conditions established by the user through the *programming options* for collecting three types of *recycled material* represented by circles *(yellow, blue, green)*. The robots colors are used to identify the *role* in the group that determines the *material* they should collect. In a typical situation of the software, the robot agent is assigned to *collect an object* and take it to the *recycling store* assigned. During this process, every agent mainly develops the processes of: *Communication*, considering that it has to report his state continuously and the presence of an additional recyclable object found in the environment; *Coordination*, during the navigation activities and taking turn to get in the recycling store whenever two or more agents with the same role wanted to do it at the same time; *Conflict resolution*, two or more robots want to share the same navigation space and when they happen to meet in the arriving time at the recycling store; *Collaborative*, inasmuch as robots collect all the recyclable objects and clean up the environment.

2.1 The Software Structure

To build this software the following technological free tools: Eclipse Kepler as IDE, Processing version 2.1 and the library AI for 2D Games and G4P to the management of robots states and the user interface creation respectively. Due to the environment design in mind for 10 and 12 years old kids, the robots programming and interaction conditions are oriented to the commands exploration which allows that through the buttons the cooperative characteristics that regulates the robots behavior could be modified. It is important to point out that for navigation, robots use artificial intelligence in order to interact with the environment in actions such as: randomly navigating, avoiding obstacles, getting close and away from an object, moving into a position between two robots, predicting positions, generating cohesion, separating and aligning with other robots.

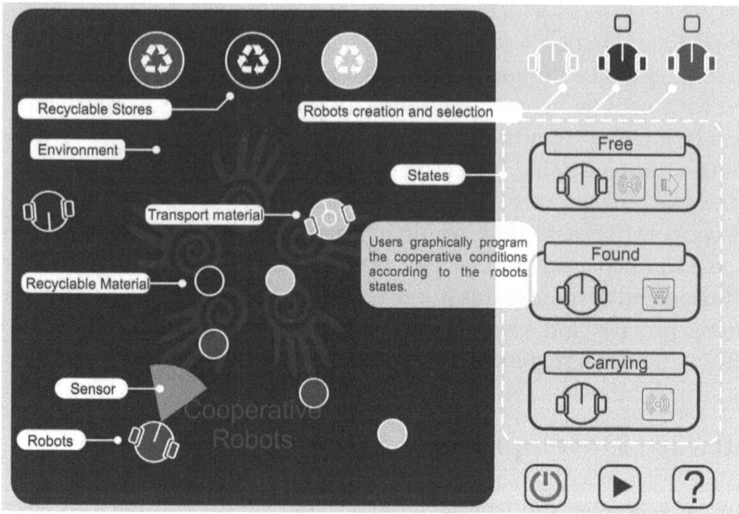

Fig. 1. Interface

The design is based on the programming paradigm of the software oriented towards events where the software flow is not sequential but it fits to the presented states on its execution. Therefore, at the beginning of those states their respective transitions are identified. The following explanation describes the possible robot state: *Global*: This state is in constant execution and it is to assign objects of recyclable material to the robots. *Follow*: Once recyclable material has been assigned to the robot it changes its state to go get it. *Load*: When the robot reaches its objective, it changes its state to load it. *Transport*: After loading the recyclable material the robot changes to this state to take it to the respective recycling store. *Deposit*: When the robot arrives to the recycling store its state changes to deposit it there. *Navigate*: Once the robot deposits the recyclable material, and if it does not have any other objects of the same kind assigned, it changes its state to navigate randomly through the environment; otherwise it changes to the *Follow* state. *Wait*: This state is executed every time a robot is addressed to its assigned recycling store and meets another robot occupying the same store; therefore it has to wait until that robot comes out to get the recyclable material in the store.

When the user creates a robot, three *recyclable material* objects from the same category are also created; the maximum amount of robots per each category is three, this is to not saturate the environment and help the robots cooperation be more fluent and comprehensible during their environment cleaning task. Once the desired amount of robots from the three categories has been created and the characteristics of its individual and cooperative behavior has been set up the user clicks on the *start* button, this leads into the *Global* state which executes the algorithm that assigns recyclable material to the robots, taking into account the Euclidian distance (meaning the closest first) that separates it from the material.

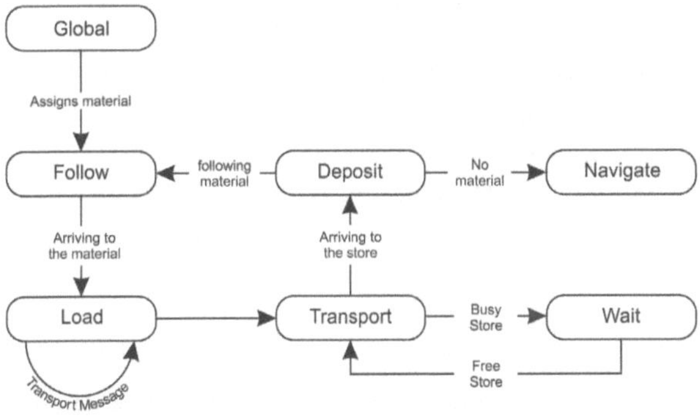

Fig. 2. Transition States Diagram

To prevent the same object of recyclable material from being assigned to two or more robots simultaneously, a Boolean variable associated to that object changes its value once this has been assigned to another robot. After the assignment, the robot changes to the *Follow* state evaluating in every cycle if it is close enough to the recyclable material, when the proximity condition is accomplished it changes to the *Load* state in which the first thing the robot does is to send a message that sets off the transition of the *Transport* state. In the last state, the robot evaluates in every cycle if it arrives to its assigned recyclable store, if so it changes the value of one Boolean variable to let all the robots know it is busy, then it moves to the middle of the store and releases the material it is transporting. In that case, it comes out from the store and it changes a Boolean variable value once again so other robots willing to get in, and are in Wait state, can move in. Once the robot is free, it verifies if it has more recyclable material assigned. If so, it changes into the *Follow* state, otherwise the robot changes into Navigate state, to move randomly along the environment. Figure 2 shows the transition states diagram already described; the states are represented by rectangles and the arrows symbolize the transition cause between those states. For the robots, state transition the library AIfor2DGames was used. With which is possible to describe a state in a simple way using three actions: *Enter*: It is used to start the entity properties. *Update*: It allows describing the behavior that an entity should have according to determined conditions. An action could turn on to be a state change. *Exit*: It is used to set up the properties that the entity should have right before the state change. The communication process, which allows the states transition, is implemented through the creation, delivery, receipt of telegrams compounded by an emitter, receptor, message identifier and a waiting time frame to be send. For the software development, the robots organization with defined specialization levels are evident in the task assignment they have to identify, to collect and to deposit one of the three kinds of recyclable materials. During its development, they transfer the state data, synchronize the actions, task are delivered, and resources are distributed. The quantity of agents serves the strategy adopted by the user; the task is accomplished by an agent which confirms there are two levels. Robots are not designed to crash, they have a function to represent their environment, to know their currently position related to the world, the other robots, and the three types of recyclable materials; they have a productive and conative function dealed in the previous explanation of the state's transition and vegetative function. This is related to the resources and organization, which highly depends on the conditions established by the user.

3 Educational Proposal

In the educative field, technology is not just a tool that helps learning processes, but it is part of the individual and distributed cognitive processes. This should be a variable in the multiple intelligences recognition[12] that explains the individual training. That reason could be considered from the *Posthumanism* conceptions[13], where man is defined as a natural and artificial system and the agent' theory is an option to explain this integration. The developed software is a cognitive activation tool that stimulates thinking actions related to the robots' cooperative work and the consequences for its use in daily basis tasks. It can be defined as an object in which the learning processes is promoted through the discovery of the robots behavior consequences provoked by the cooperative conditions established by the user. An advantage during the learning process is the real time correction that the user can make when their hypothesis in programming is not efficient to the development of the cooperative task. During the robot programming process, the user understands that the cooperative work is not just about the people gathering, but life itself is a cooperative process with other persons, where the group and individual identity establishes communication and a cooperative relationship. While the constant conflicts resolution either positively or negatively. In fact, throughout the robots programming and the complexity increase, cooperation processes are reinforced, which are understood as a cooperation result. To achieve this process it is necessary for the individuals to be motivated through the acknowledgement of the award factors, to identify if their aims are related to the other members' objectives, and to be aware about while accomplishing their own aims, they are helping others'; then they will be ready to share information, to exchange resources, and to bring help and support[14].

4 Software Evaluation

The characteristics of the final software version are originated on a previous version validated in 12 public and private institutions from Bogotá City with fifth and sixth graders. As a result of this stage, some changes in the agents' cooperative methodology and in the users' programming conditions were made. However, another result of this stage was the complex system representation in dynamic virtual environments such as the cooperative work in social organizations, facilitating the process, and take ownership of information of the students due to the fast feedback process[15]. Besides the users' motivation was originated by the fast conceptual change that was evident, because the software is not a traditional tool in the classroom. Realizing that, it can be learned in a personalized learning process. The second validation process was applied to students of 11 public and private institutions from Bogotá and Santa Marta Cities. The intervention strategy started with an introduction of evidence and cooperative meaning in the living beings; it continued with the key cooperative concepts according to the individuals context. The virtual environment was presented after for its exploration where the user programmed the robots. Finally, a semi-structure interview was performed to identify the software relevancy to benefit the cooperative work.

5 Conclusions and Further Works

Through the first interviews done to one of the 184 kids group, it is easy to find that the initial motivation of the cooperative work, which is external, commonly is done by the teacher; this demonstrated the lack of conscience about the benefits of cooperative work. After applying the software, they recognized all those benefits and comprehended its dynamical development not only in the recycling concerned but in other basis daily activities. The collected data during the experimental process proved how the real time feedback allowed the users to realize the conceptual errors they had made when they set up the cooperative task development, especially referring to: coordination, communication, and conflict resolution. In the educative field, the cooperative robots programming establishes a relationship between cognitive processes related to cooperation, the task, and the software. In the literature review, it was evident that multiagents systems are an alternative, which contributes significantly in the artificial environment and posthumanism construction. The developed software is a cognitive activation tool that reflects accurately the basic social behavioral processes similar to the ones the users will face throughout his or her life. Once the field work experience was finished and before introducing the game dynamic, it was evident that there was not clarity in the concepts that referred to communication, coordination, cooperation, and conflict resolution, which makes context necessary with daily basis tasks that kids do; among them related to academic and non academic aspects of their lives, after getting close to this concepts, they get close to the software, they show interested about it and it becomes a new game for them, subsequent to a basic explanation about the game working, they easily found out the functions and objectives of each robot. Afterwards, questions about the importance of cooperative work, the software use as a tool to learn, and cooperate and the cooperative importance in group tasks were asked. At this moment, it is easy to see kids do recognize the communication, cooperation, and conflict resolution concepts and their importance of all of this after they use the software, concluding that *"If they cooperate to each other and they know how to communicate, everyone will accomplish the objectives faster by getting the solution to the established problem. Besides, with an accurate communica-*

tion, it is easy to solve conflicts, achieve group goals, and solve problems from the beginning in order to get a good result".

References

1. Manuel, A.S.V., Carlos, G.C.: Trabajo en Equipo, el caso colombiano. Red de Revistas Científicas de América Latina, el Caribe, España y Portugal. Análisis Económico XX(43), 147–165 (2005)
2. Colombia, MEN. Estándares básicos de competencias ciudadanas, Colombia (2004)
3. Gregorio, C., Germán, C.: Investigación en administración en Ámerica Latina: Evolución y Resultados. Universidad Nacional. Colombia (2005)
4. Halverson, C.B., Aqeel, T.: Effective Multicultural Teams: Theory and Practice. Springer Science (2008)
5. West, M.A., Tjosvold, D., Smith, K.G.: International Handbook of Organizational Teamwork and Cooperative Working, England (2003)
6. Len, F.: The Perfect Swarm, The Science of complexity in everyday life. Basic Book, New York (2009)
7. Cao, Y.U., Fukunaga, A.S., Kahng, A.B.: Cooperative Mobile Robotics: Antecedents and Directions (1997)
8. Jacques, F.: Multi-Agent System: An Introduction to Distributed Artificial Intelligence. Addison Wesley Longman, Harlow (1999)
9. Lewis, M.: Technics and Civilization. University of Chicago Press Edition (2010)
10. Brown, L.A., Walker, W.H.: Prologue: Archaeology, Animism and Non-Human Agents. Journal of Archaeological Method and Theory, 297–299 (2008)
11. Deval, J.: El animismo y el pensamiento infantil. Siglo XXI Editores, Madrid (1975)
12. Howard, G.: Frames of Mind, The Theory of Multiple Intelligences. Basic Books (1993)
13. Katherine, H.: How We Became Posthuman, Virtual Bodies in Cybernetics, Literature, and Informatics. The University of Chicago Press, Chicago er London (1999)
14. Michael, W., Dean, T., Ken, S.: International Handbook of Organizational Teamwork and Cooperative Working. John Wiley and Sons Ltd. (2003)
15. Wheeler, S.M.: Role Playing Games and Simulations for International Issues Courses. Journal of Political Science Education 2(3), 331–347 (2006)
16. Margaret, H., Margaret, H.: Learning Science Through Computer Games and Simulations. Committee on Science Learning: Computer Games, Simulations, and Education; National Research Council (2011)
17. Nass, C., Moon, Y.: Machines and mindlessness: Social responses to computers. Journal of Social Issues 56(1), 81–103 (2000)
18. Ron, S.: Cognition and Multi-Agent Interaction From Cognitive Modeling to Social Simulation. Cambridge University Press (2006)
19. Informe nacional sobre desarrollo humano 2013. Programa de las Naciones Unidas para el Desarrollo PNUD. 325 Páginas (2013)
20. Mitnik, R., Recabarren, M., Nussbaum, M., Soto, A.: Collaborative robotic instruction: A graph teaching experience (2009)
21. Barella, A., Valero, S., Carrascosa, C.: JGOMAS: New Approach to AI Teaching. IEEE Transactions on Education 52(2), 228–235 (2009)

A Team Formation Tool for Educational Environments

Elena del Val, Juan Miguel Alberola, Victor Sanchez-Anguix,
Alberto Palomares, and Mª Dolores Teruel

Departament de Sistemes Informàtics i Computació,
Universitat Politècnica de València,
Camí de Vera s/n, 46022, València, Spain
{edelval,jalberola,sanguix,apalomares}@dsic.upv.es,
dteruel@upvnet.upv.es

Abstract. Teamwork is a critical competence in the higher education area, and it has become a critical task in educational and management environments. Unfortunately, looking for optimal or near optimal teams is a costly task for humans due to the exponential number of outcomes. In this paper, we present a web application that facilitates the task of automatic team generation of near optimal teams based on collective intelligence, coalition structure generation, and Bayesian learning. This tool has been used in real classroom scenario and the data collected from the experience has been used as input for synthetic simulations. The experiments show that the tool is able of converging towards the optimal solution (team formation) as long as students do not have great difficulties evaluating others.

Keywords: Team formation, Coalitions, Education.

1 Introduction

Nowadays, it is widely accepted that teamwork skills are at the heart of organizations' success [20,4,22]. In fact, many complex projects are carried out by multidisciplinary teams whose individuals come from different academic backgrounds. In these situations, cooperation mechanisms and correct team dynamics are necessary to aggregate the wide expertise of team members, and to accomplish the team's goal. In fact, team dynamics have been reported to be of critical importance to teams' success [5,13,2]. Inadequate team dynamics can lead to opposite results, even if individuals who form the team are successful workers. For all of these reasons, there has been a wide interest in the application of teamwork skills at the classroom. Firstly, it allows students to improve their teamwork skills in a scenario that is similar to the organizational world. Secondly, teamwork is highly correlated with cooperative learning, an active learning methodology that has been documented to report more positive learning results than classic methodologies [9,8]. Due to these reasons, the European Higher Education Area has described teamwork as a competence on the core of its new education policies [10,12].

One of the main theories for forming successful teams is Belbin's role taxonomy [1]. In that work, Belbin identifies nine behavioral patterns (i.e., roles) that help teams to succeed in their task: *plant, resources investigator, co-ordinator, shaper, monitor evaluator, teamworker, implementer, and completer finisher*. One of the key concepts underlying this theory is that teams should be formed by individuals playing heterogeneous

J.B. Perez et al. (eds.), *Trends in Practical Applications of Heterogeneous Multi-agent Systems. The PAAMS Collection*, Advances in Intelligent Systems and Computing 293,
DOI: 10.1007/978-3-319-07476-4_21,

roles [1,7]. Therefore, in order to guarantee proper cooperation in a team, including teams of students, teams should be formed attending to the characteristics of individuals. The impact of Belbin's taxonomy has given birth to a wide variety of studies that attempt to form optimal teams of students according to each individual's prominent Belbin role [3,6,15]. Many of these approaches usually rely on the completion of a self-assessment form by students that determines their prominent role. However, in many occasions, one's own perception may differ from the actual behavioral patterns shown at the classroom. This situation is also commented by different scholars [19,21].

Since other students' observations may be a more accurate source of information for estimating the prominent Belbin role teammates, aggregating the opinion provided by students becomes a useful mechanism for determining Belbin's roles. Nevertheless, one should consider that unanimously agreeing on any matter is a very unlikely situation, and, therefore, uncertainty arises as student express different opinions. One of the many applications of Artificial Intelligence is tacking decisions in uncertain environments [17]. On top of that, the problem of forming optimal groups of individuals, known in the literature as coalition structure generation [18,14], has an exponential cost that exceeds the cognitive capabilities of humans. One of the many applications of Artificial Intelligence is taking optimal, or near optimal solutions, for exponential problems that go beyond the calculation capabilities of humans. Summing up both arguments, Artificial Intelligence becomes a perfect candidate for the problem of forming optimal teams of students in the classroom.

In this paper we present a tool that allows to form teams of students according to Belbin's role taxonomy. It employs Bayesian learning [17] in order to tackle uncertainty regarding students' roles, and coalition structure generation mechanisms for finding optimal teams according to the information provided by students. The rest of this paper is organized as follows. First we present the theoretical an practical model underlying our tool in Section 2. Then we introduce a case of study based on an ongoing experiment carried out in a university course. Finally, we briefly explain the conclusions of this paper and we highlight some future lines of work.

2 Team Formation Tool

During a course, the teacher carries out several team activities in the classroom. The main problem of teachers is how to create teams when there is a high number of students and there is no previous information about the profiles of the students. Considering a classroom of 60 students to be grouped in teams of 6, over than 50 millions of teams can be obtained. In order to support the teacher with a framework to create and manage teams, we propose a software application that prevents teachers from carrying out the costly task of dividing students into optimal or near optimal teams. The application relies on collective intelligence, coalition formation, and Bayesian learning to form proper distributions of students teams.

The proposed application uses a web platform where the actors, teachers and students, can interact with the system. The main functionalities of the system can be observed in the UML diagram in Figure 1. Both, teachers and students, must log in in the system in order to start to interact with the application. This login allows the system to show a personalized view of the possible actions available for each role.

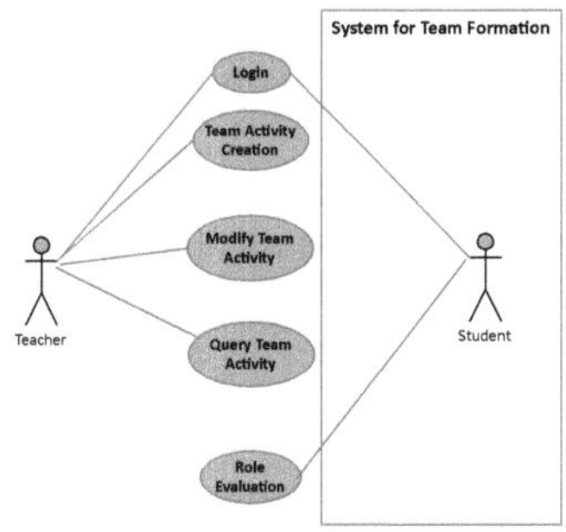

Fig. 1. Use case for the Team Formation Application

The usual workflow of the application starts with the creation of a new team activity for an specific course. This action has as result a list of students teams for the team activity. If there is no previous information about the students in the system database, the system provides a random set of groups of students. Otherwise, the system provides a set of groups based on previous evaluations. In the case that the teacher does not agree with the data about the activity or the proposed teams, he/she can modify them through the action *Modify Team Activity*. When teacher considers appropriate, he/she can publish the teams that are notified to the students through an email.

Once the activity has been carried out, the students are allowed to provide their opinion about their teammates in the activity. Basically, each team member, gives an opinion about which is the most predominant role in each teammate. The result of this process is that the new information is stored in a database to be considered in future team formations. This set of actions is repeated during the team activities in the course. The idea behind this iterative sequence of actions is that as the system obtains more information, the system will have more evidences about the predominant roles in each student. This fact should allow the system to form more adequate teams in each iteration.

In the following sections, we describe with detail the two main actions of the web application: *Team Activity Creation* and *Role Evaluation*.

2.1 Team Activity Creation

At the start, the teacher should login in the system (see Figure 2), and then he can create a new activity. After that, the web shows a pull-down menu with the courses associated to the teacher. The teacher selects the activity where he/she is going to create the team activity. Optionally, he/she can fill out all the fields that describe the activity (activity description, start date, end date, on-line material for the activity). The next step consists on determining the maximum and minimum size for the students teams. With these parameters, the application is ready to generate an automatic proposal of teams. If there is no previous information about students, the first set of teams is generated randomly. Otherwise, the system based on the information from previous activities uses a coalition algorithm to generate the set of teams for the activity.

To calculate the teams for the next activity, the system is based on a process of structural coalition formation. The problem of coalition formation can be described as follows.

Let $A = \{a_i, ..., a_n\}$, be a set of students, and $R = \{r_1, r_2, ..., r_m\}$ be the set of roles that the student can play (in our case it is the set of Belbin's roles), and let $role_i$ denote the true predominant role of a_i.

A subset $T \in A$ is called a *team*, and a *team structure* $S = \{T_1, T_2, ..., T_k\}$ is a partition of disjoint teams such that $\bigcup_{\forall T_j \in S} T_j = A$ and $S \in 2^A$. The goal of the application is determining an optimal team structure for the classroom $\underset{S \in 2^A}{argmax}\ v(S)$, where $v(S)$ is a evaluation function for the team structure. In this study, we will consider that the quality of each team is independent of other teams. Hence, we can calculate the value of the team structure as $v(S) = \sum_{T_j \in S} v(T_j)$. The value of a team $v(T_j)$ can be calculated attending to the predominant role that each student $a_i \in T_j$ has $(role(a_i))$. Let $|T_j| = k$ denote the size of the team and $\pi_j = \{r'_1, ..., r'_k\}$ with $\forall r'_i \in R$ be a vector with the true predominant role of each team member. In that case, $v(T_j) = v(\pi_j)$. According to different studies [7], the team should benefit from having a balanced distributions of roles (i.e., one person per role). This score can be provided by an expert.

Unfortunately, it is not possible to accurately know the predominant role of each team member π_j and therefore $v(\pi_j)$ cannot be calculated with precision. However, it is possible for us to calculate an estimation of the value of the coalition given the history of evaluations H that is gathered from the students during the course. Let $\pi' = \{role_1 = r'_1, ..., role_k = r'_k\}$ be a vector containing a set of hypotheses for the predominant roles of each team member, and Π be the set of all possible vectors of hypotheses for predominant roles of T_j. In that case, we can calculate the expected value of a team given the history of evaluations as:

$$\hat{v}(T_j|H) = \sum_{\pi' \in \Pi} p(\pi'|H) \times v(\pi') = \sum_{\pi' \in \Pi} v(\pi') \times \prod_{a_i \in T_j} p(role_i = r'_i|H) \qquad (1)$$

where and $p(\pi'|H)$ represents the probability for π' to be the real role distribution in T_j given the history of evaluations H. Each $p(\pi'|H)$ can be divided into its $p(role_i = r'_i|H)$ since we assume that the role of each student is conditionally independent given the history of evaluations. Therefore, our team formation problem at each iteration is casted out to one problem that follows the next expression:

$$\underset{S \in 2^A}{argmax} \sum_{T \in S} \hat{v}(T|H) \qquad (2)$$

It turns out that partitioning a set students into disjoint teams while optimizing a social welfare function corresponds to the formalization of coalition structure generation problems. For our simulation experiments, we formalize the coalition structure generation problem as a linear programming problem [11] and solve it with the commercial software *ILOG CPLEX 12.5* [1].

[1] http://www.ibm.com/software/commerce/optimization/cplex-optimizer/

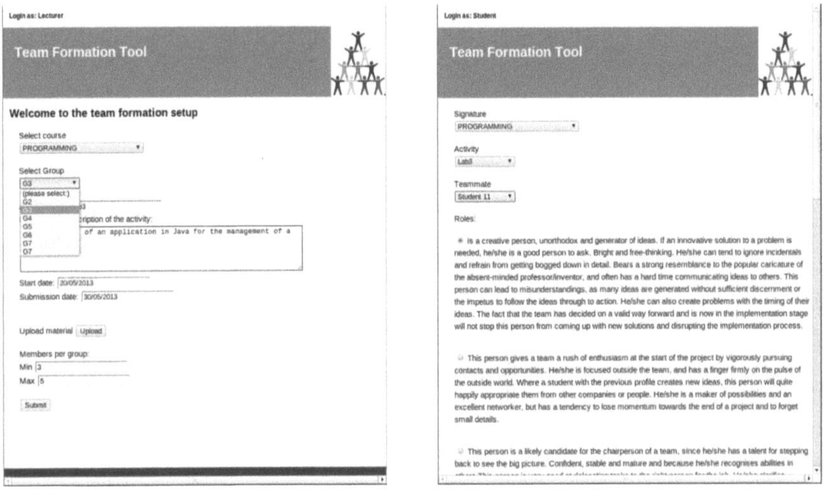

Fig. 2. (Left) Form to introduce the information about the team activity. (Right) Form to evaluate teammates.

2.2 Role Evaluation

Once the activity is finished, each student receives a notification from the system to evaluate his teammates according to Belbin's roles. Each student should login the software application. Then, he/she selects the course, the activity inside the course, and finally the teammate to evaluate 2. Then, the application shows a full description of Belbin's roles. The application does not show the name of the roles, only the description of the main features with the aim of avoiding skewed opinions in the students. At that point, the student should classify each teammate into a role. The information is then gathered by the application and stored in a database.

After the peer evaluation, the application updates the information for each student. Then, new information becomes available regarding the most predominant role of each student and the history of evaluations H grows. Hence, at each iteration we can update information regarding the probability for an agent a_i to have r'_i as his/her most predominant role given the evaluation history $p(role_i = r'_i|H)$. We employ Bayesian learning for this matter :

$$p(role_i = r'_i|H) = \frac{p(H|role_i = r'_i) \times p(role_i = r'_i)}{\sum\limits_{r \in R} p(H|role_i = r) \times p(role_i = r)} \tag{3}$$

where $p(H|role_i = r'_i)$ is the likelihood function and $p(role_i = r'_i)$ is the prior probability for the hypothesis. For the likelihood function, we can calculate it as $p(H|role_i = r'_i) = \frac{\#\{r'_i \in H_i\}}{|H_i|}$, where H_i denotes the peer evaluations about agent a_i, and $\#\{r'_i \in H_i\}$ indicates the number of times that r'_i appears as evaluation in H_i. As for the prior probability, we calculate it as $p(role_i = r'_i) = \frac{\#\{r'_i \in H\}}{|H|}$. Laplace smoothing [16] is employed to ensure that the likelihood for each role hypothesis can be calculated in the first iterations.

3 Case Study

In order to validate the team formation model provided by the tool, in this section we present a case study tested in the 2013-2014 course of Tourism Management at the Universitat Politècnica de València. This course is composed by 60 students that were distributed into teams.

At the beginning of the course, we did not have any information regarding the roles of the students. Therefore, the whole 60 students were firstly grouped into teams of 6 members according to a random criteria, in order to delevop a project. After finishing this project, all the students entered the information regarding their partners. It is important to remark that the evaluation of each individual after this project, would provide a general opinion due to the evaluation of those who worked which him. This general opinion determines the main role associated to this individual. Instead, if relying only on the own self-perception, individuals main role would be determined by considering the self-opinion, which may cause a biased view.

In addition, students were asked about their own self-perception by a test which comprised several parameters associated to the roles. The differences between the self-perception of individuals and the general opinion is around 70%. This means that the self-perception of each student was different from the general opinion of his partners in 7 out of every 10 students. This is a significant, because one individual may be affected by his self-awareness or aspirations to a *co-ordinator* role, but exhibits the behavior of a *plant* role. Therefore, his creative abilities are more appreciated by the others rather than his abilities as *co-ordinator*.

According to these evaluations, Figure 3 shows the number of students which were evaluated by their partners according to these main roles. As it can be observed, the distribution of roles seems to be heterogenous. As an example, it can be appreciated that the number of *co-ordinators* in the whole team is much lower than the number of *plants*. This is remarkable, since the number of students who correspond to a *co-ordinator* role according to their own self-perception is around the 20% of students. In contrast, this percentage is slightly higher than the 3% according to the opinion of the partners.

We must point that some students were not matched to any main role (e.g. an student who was evaluated as four different roles by four partners). However, considering that this is the first project, it is assumed that when performing more projects, the number of evaluations will increase and the roles could be defined more accurately.

Although the number of evaluations for each member is higher than only considering the self-perception of each individual, increasing the number of evaluations through several projects would also increase the significance of the roles associated. According to Belbin, a team composed by heterogenous roles would provide higher effectiveness than a team composed by homogeneous roles. Therefore, we can measure this heterogeneity depending on the roles distribution. As a result, a team composed by heterogenous roles has associated a heterogeneity of 1, while a team composed by homogeneous

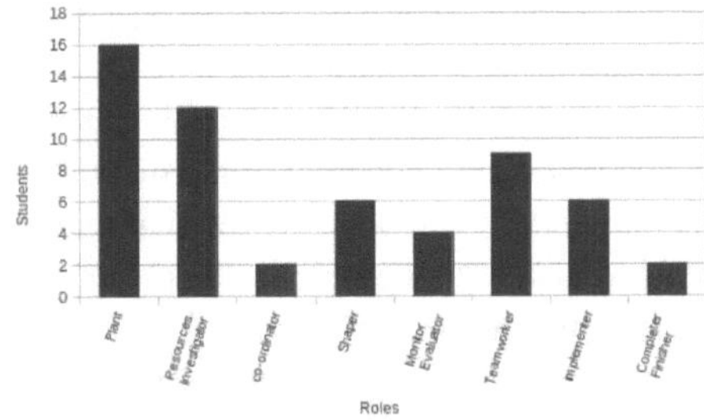

Fig. 3. Roles distribution

role has a minimum heterogeneity of 0.125. This heterogeneity is related to the satisfaction of the students when working together.

3.1 Simulation

Assuming that the distribution of roles follows the distribution shown in Figure 3, we can simulate the performance of the teams, according to the algorithm provided by the team formation tool (Section 2.1. To do this, we start from the initial distribution of teams and perform 5 iterations of the algorithm (with would correspond with the forthcoming projects). We must note that the distribution of roles is not known by the students, who are only responsible of providing evaluations after each project.

Figure 4 shows the heterogeneity of the teams for each project. We test two different scenarios, one in which the roles of the students are strongly defined and thus, the students are able to determine the main role easier, and other scenario in which the roles of the students are weakly defined and thus, it is more difficult to determine the main role.

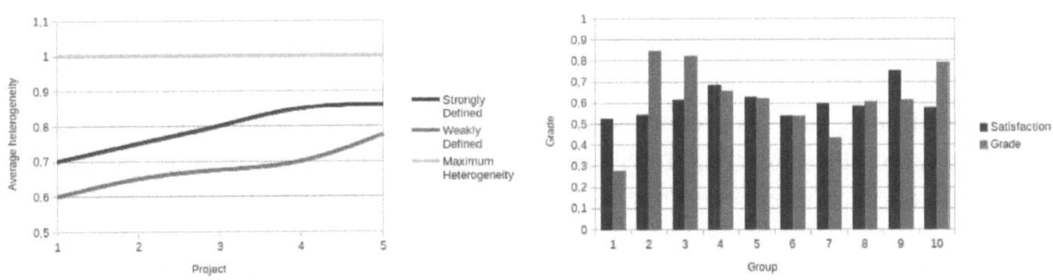

(a) Execution.

(b) Notes.

Fig. 4. Execution and Notes results

Although the distribution of roles in a real scenario is not homogeneous (there are more students whose main role is role 1 than role 3), it can be observed that as the number of projects increase (and therefore, the evaluations of students), the more heterogenous teams are found after each project. It can be observed that in case that roles are strongly defined (e.g. a student who clearly corresponds to an specific role), the heterogeneity increases earlier.

Figure 4b shows the average satisfaction of each team according to the evaluation of their members, with the real grade obtained with this project. Apparently, there is some relationship with both parameters. It can be observed that teams with high satisfaction (e.g. teams 4 and 9) obtained also a good grade in the project, while teams with lower satisfaction (e.g. team 1) obtained a low grade.

As the feedback provided by the students, we should point out that although the students initially did not like the grouping methodology (since they preferred to work with their friends), they were still able to work together as a team. As a future work, we expect to perform several projects in order to improve the accuracy of the estimation and therefore, the teams performance.

4 Conclusions

In this paper we have presented a tool that facilitates the task of generating teams of students in an educational context. The tool is a web application based on Belbin's role taxonomy [1], collective intelligence, coalition structure generation algorithms, and Bayesian learning. The application facilitates the interaction of the teacher and students with the system. Basically, there are two main stages in the workflow of the application. In a first stage, the teacher generates an activity and the tool recommends a set of teams based on an algorithm of coalition formation. There is a second stage where the students provide their opinion about which is the most predominant role in each teammate. This information is used to improve the team formation process for next activities through bayesian learning. We have tested this tool in a real context and we have used the data collected as input for synthetic simulations. The results show that as long as students do not have great difficulties classifying others, the policy is capable of improving the quality of team structures in a few iterations and gradually converging towards the optimal solution.

We simulated different scenarios in order to test different environmental conditions. The results are encouraging enough to continue this research. As a future work, we plan to extend the experiments in order to consider large populations of students and environmental conditions, such as scenarios where some roles are more important than others. In addition, we also intend to study whether or not the inclusion of more attributes in the classification problem can improve the performance of the policy.

Acknowledgements. This work is supported by TIN2011-27652-C03-01 and TIN2012-36586-C03-01 of the Spanish government.

References

1. Belbin, R.M.: Team roles at work. Routledge (2010)
2. Colomo-Palacios, R., Casado-Lumbreras, C., Soto-Acosta, P., Garcia-Peñalvo, F.J., Tovar, E.: Project managers in global software development teams: a study of the effects on productivity and performance. Software Quality Journal, 1–17 (2012)

3. Furnham, A., Steele, H., Pendleton, D.: A psychometric assessment of the belbin team-role self-perception inventory. Journal of Occupational and Organizational Psychology 66(3), 245–257 (1993)
4. Guzzo, R.A., Dickson, M.W.: Teams in organizations: Recent research on performance and effectiveness. Annual Review of Psychology 47(1), 307–338 (1996)
5. Guzzo, R.A., Shea, G.P.: Group performance and intergroup relations in organizations. In: Handbook of Industrial and Organizational Psychology, vol. 3, pp. 269–313 (1992)
6. Henry, S.M., Todd Stevens, K.: Using belbin's leadership role to improve team effectiveness: an empirical investigation. Journal of Systems and Software 44(3), 241–250 (1999)
7. Higgs, M., Pewina, U., Ploch, J.: Influence of team composition and task complexity on team performance. Team Performance Management 11(7/8), 227–250 (2005)
8. Johnson, D.W., Johnson, R.T.: An educational psychology success story: Social interdependence theory and cooperative learning. Educational Researcher 38(5), 365–379 (2009)
9. Johnson, D.W., Johnson, R.T., Smith, K.: The state of cooperative learning in postsecondary and professional settings. Educational Psychology Review 19(1), 15–29 (2007)
10. Maffioli, F., Augusti, G.: Tuning engineering education into the european higher education orchestra. European Journal of Engineering Education 28(3), 251–273 (2003)
11. Ohta, N., Conitzer, V., Ichimura, R., Sakurai, Y., Iwasaki, A., Yokoo, M.: Coalition structure generation utilizing compact characteristic function representations. In: Gent, I.P. (ed.) CP 2009. LNCS, vol. 5732, pp. 623–638. Springer, Heidelberg (2009)
12. Pajares, S., Sanchez-Anguix, V., Torreño, A., Esparcia, S.: A novel teaching-learning strategy for teamwork based on agreement technologies. In: IJCA Proceedings on Design and Evaluation of Digital Content for Education (DEDCE), vol. (1), pp. 21–30 (2011)
13. Parker, G.M.: Team players and teamwork: New strategies for developing successful collaboration. John Wiley & Sons (2008)
14. Rahwan, T., Ramchurn, S.D., Jennings, N.R., Giovannucci, A.: An anytime algorithm for optimal coalition structure generation. Journal of Artificial Intelligence Research 34(2), 521 (2009)
15. Rajendran, M.: Analysis of team effectiveness in software development teams working on hardware and software environments using belbin self-perception inventory. Journal of Management Development 24(8), 738–753 (2005)
16. Russell, S.J., Norvig, P.: Artificial Intelligence: A Modern Approach. Prentice Hall (2010)
17. Russell, S.J., Norvig, P., Canny, J.F., Malik, J.M., Edwards, D.D.: Artificial intelligence: a modern approach, vol. 74. Prentice Hall, Englewood Cliffs (1995)
18. Sandholm, T., Larson, K., Andersson, M., Shehory, O., Tohmé, F.: Coalition structure generation with worst case guarantees. Artificial Intelligence 111(1), 209–238 (1999)
19. Senior, B., Swailes, S.: A comparison of the belbin self perception inventory and observer's assessment sheet as measures of an individual's team roles. International Journal of Selection and Assessment 6(1), 1–8 (1998)
20. Sundstrom, E., De Meuse, K.P., Futrell, D.: Work teams: Applications and effectiveness. American Psychologist 45(2), 120 (1990)
21. Swailes, S., McIntyre-Bhatty, T.: The belbin team role inventory: reinterpreting reliability estimates. Journal of Managerial Psychology 17(6), 529–536 (2002)
22. Wheelan, S.A.: Creating effective teams: A guide for members and leaders. Sage (2010)

Retraction Note to: Trends in Practical Applications of Heterogeneous Multi-agent Systems. The PAAMS Collection

Javier Bajo Perez, Juan M. Corchado Rodríguez, Philippe Mathieu, Andrew Campbell, Alfonso Ortega, Emmanuel Adam, Elena M. Navarro, Sebastian Ahrndt, María N. Moreno, and Vicente Julián

Retraction Note to: Chapter 4 in: J.B. Perez et al. (eds.), *Trends in Practical Applications of Heterogeneous Multi-agent Systems*. The PAAMS Collection, Advances in Intelligent Systems and Computing 293, https://doi.org/10.1007/978-3-319-07476-4_4

The Series Editor and the publisher have retracted this chapter. An investigation by the publisher found that a number of chapters, including this one, from multiple conference proceedings raise various concerns, including but not limited to compromised editorial handling, inappropriate or unusual citation behavior and undisclosed competing interests. Based on the findings of the investigation, the Series Editor and the publisher no longer have confidence in the results and conclusions of this chapter.

Authors Ana Belen Gil and Tiancheng Li disagree with this retraction. Author Álvaro Lozano has not responded to correspondence regarding this retraction.

The retracted version of these chapters can be found at
https://doi.org/10.1007/978-3-319-07476-4_4
https://doi.org/10.1007/978-3-319-07476-4_5
https://doi.org/10.1007/978-3-319-07476-4_7
https://doi.org/10.1007/978-3-319-07476-4_14
https://doi.org/10.1007/978-3-319-07476-4_15

J.B. Perez et al. (eds.), *Trends in Practical Applications of Heterogeneous Multi-agent Systems. The PAAMS Collection*, Advances in Intelligent Systems and Computing 293,
https://doi.org/10.1007/978-3-319-07476-4_22,

Retraction Note to: Chapter 5 in: J.B. Perez et al. (eds.), *Trends in Practical Applications of Heterogeneous Multi-agent Systems.* The PAAMS Collection, Advances in Intelligent Systems and Computing 293, https://doi.org/10.1007/978-3-319-07476-4_5

The Series Editor and the publisher have retracted this chapter. An investigation by the publisher found that a number of chapters, including this one, from multiple conference proceedings raise various concerns, including but not limited to compromised editorial handling, inappropriate or unusual citation behavior and undisclosed competing interests. Based on the findings of the investigation, the Series Editor and the publisher no longer have confidence in the results and conclusions of this chapter.

Authors Daniel Hernández and Ana de Luis have not responded to correspondence regarding this retraction. The Publisher has not been able to obtain a current email address for author Sigeru Omatu.

Retraction Note to: Chapter 7 in: J.B. Perez et al. (eds.), *Trends in Practical Applications of Heterogeneous Multi-agent Systems.* The PAAMS Collection, Advances in Intelligent Systems and Computing 293, https://doi.org/10.1007/978-3-319-07476-4_7

The Series Editor and the publisher have retracted this chapter. An investigation by the publisher found that a number of chapters, including this one, from multiple conference proceedings raise various concerns, including but not limited to compromised editorial handling, inappropriate or unusual citation behavior and undisclosed competing interests. Based on the findings of the investigation, the Series Editor and the publisher no longer have confidence in the results and conclusions of this chapter.

Authors Amparo Casado, Amparo Jiménez and Javier Bajo have not responded to correspondence regarding this retraction. The Publisher has not been able to obtain a current email address for author Sigeru Omatu.

Retraction Note to: Chapter 14 in: J.B. Perez et al. (eds.), *Trends in Practical Applications of Heterogeneous Multi-agent Systems*. The PAAMS Collection, Advances in Intelligent Systems and Computing 293, https://doi.org/10.1007/978-3-319-07476-4_14

The Series Editor and the publisher have retracted this chapter. An investigation by the publisher found that a number of chapters, including this one, from multiple conference proceedings raise various concerns, including but not limited to compromised editorial handling, inappropriate or unusual citation behavior and undisclosed competing interests. Based on the findings of the investigation, the Series Editor and the publisher no longer have confidence in the results and conclusions of this chapter.

The authors have not responded to correspondence regarding this retraction.

Retraction Note to: Chapter 15 in: J.B. Perez et al. (eds.), *Trends in Practical Applications of Heterogeneous Multi-agent Systems*. The PAAMS Collection, Advances in Intelligent Systems and Computing 293, https://doi.org/10.1007/978-3-319-07476-4_15

The Series Editor and the publisher have retracted this chapter. An investigation by the publisher found that a number of chapters, including this one, from multiple conference proceedings raise various concerns, including but not limited to compromised editorial handling, inappropriate or unusual citation behavior and undisclosed competing interests. Based on the findings of the investigation, the Series Editor and the publisher no longer have confidence in the results and conclusions of this chapter.

Authors Daniel Hernández and María Moreno have not responded to correspondence regarding this retraction. The Publisher has not been able to obtain a current email address for author Sigeru Omatu.

Author Index

MIX
Papier aus verantwortungsvollen Quellen
Paper from responsible sources
FSC® C105338

If you have any concerns about our products,
you can contact us on
ProductSafety@springernature.com

In case Publisher is established outside the EU,
the EU authorized representative is:
Springer Nature Customer Service Center GmbH
Europaplatz 3, 69115 Heidelberg, Germany

Printed by Libri Plureos GmbH
in Hamburg, Germany